World Heritage Canal

World Heritage Canal

Thomas Telford and the Pontcysyllte Aqueduct

Paul A. Lynn

Whittles Publishing

Published by
Whittles Publishing Ltd.,
Dunbeath,
Caithness, KW6 6EG,
Scotland, UK

www.whittlespublishing.com

ISBN 978-184995-398-6

front cover: The Institution of Civil Engineers (Telford portrait);
Skyscan Photolibrary/Alamy Stock Photo (Canal boat
crossing Pontcysyllte Aqueduct Llangollen Wales)
back cover: Wikipedia/Akke Monasso

Contents

Preface

Until recently Pontcysyllte, one of the greatest aqueducts built since Roman times, was hardly known outside Wales and the ranks of canal enthusiasts. Its elevation to UNESCO World Heritage status in 2009 changed all that, and has led to heightened public interest in its designer, the great engineer Thomas Telford.

Telford has long been celebrated in Britain for his iconic Menai Suspension Bridge in North Wales, opened in 1826 and fully exposed to public view as part of his great road that stretched from London to Holyhead. Pontcysyllte Aqueduct, completed 21 years earlier, represented an equally daring advance in civil engineering, but was designed for canal boats rather than wheeled vehicles, and was tucked away in a relatively unfrequented valley. Following a rapturous opening ceremony and initial commercial success, a gradual decline of the canal system from about 1840 onwards, resulting from the rapid spread of the new railway companies, made the Llangollen Canal look increasingly like a rural backwater. Pontcysyllte remained largely ignored by officialdom, and indeed by the public at large, for the next 100 years.

Today, however, there is great interest in our historic transport infrastructure. The 19th-century engineers who did so much to pioneer and improve it are rightly seen as heroes. We realise how much we owe Telford and others for creations that have stood the test of time, and marvel at their courage and daring in an age when construction projects – even major ones such as this – relied heavily on pickaxes, wheelbarrows and human muscle power.

My own interest in the Pontcysyllte Aqueduct springs from a love of narrowboats and the canal voyages my wife and I enjoyed over a period of 25 years. Cruising the Llangollen Canal and crossing Pontcysyllte was the high point of our explorations, and has led me to a lasting fascination with the life and times of Thomas Telford, one of the world's greatest civil engineers.

Paul A. Lynn

Acknowledgements

I am grateful to the following organisations and individuals for images which do so much to bring this story to life. All are acknowledged where they appear in the text.

Institution of Civil Engineers (ICE), London

Royal Commission on the Ancient and Historic Monuments of Wales (RCAHMW), Aberystwyth

Jessica Armstrong

Richard Turner-Thomas

And the following *Geograph* photographers:

James Allan	Chris Allen	Chris Andrews
James Barton	Walter Baxter	Andy Beecroft
Humphrey Bolton	Jeremy Bolwell	Jeff Buck
Ian Capper	John Chroston	Richard Croft
Steve Daniels	Perry Dark	David Dixon
Oliver Dixon	Eirian Evans	Mark Evison
Rob Farrow	Chris Gunns	Philip Halling
Bob Harvey	John Haynes	John Horton
Christine Johnstone	Roger Kidd	John Lord
Stephen McKay	Kevin Skidmore	Colin Smith
David Smith	Jo Turner	Star Watcher
John M. Wheatley	John Winder	

Special thanks are due to Roger Kidd for a fine set of photographs taken along the Llangollen Canal.

Among the many books I have consulted, Anthony Burton's *Thomas Telford: Master Builder of Roads and Canals* and Richard Turner-Thomas's *Pontcysyllte Aqueduct and Restoration* have been especially helpful.

Finally, there are three classic books by L.T.C. (Tom) Rolt that introduced me to the English canals: *Narrowboat* (1944); *Thomas Telford* (1958); and *Landscape with Canals* (1977). Without them my wife, Ulrike, and I would never have owned a narrowboat, let alone traversed the Pontcysyllte Aqueduct. I am very grateful to Tim Rolt for permission to include a number of quotations from his father's inspirational writing.

Paul A. Lynn

1 Historic aqueducts

We had come a long way at 4 miles per hour and so could hardly turn back. The long-imagined silver stream stretched ahead, narrow as an arrow, disturbingly high above the river valley. Walkers were offered a generous railing on one side and, at worst, a ducking on the other; but we had to coax our 50-foot narrowboat along a trough with an unprotected outside edge – a tricky prospect for a skipper with a dubious head for heights.

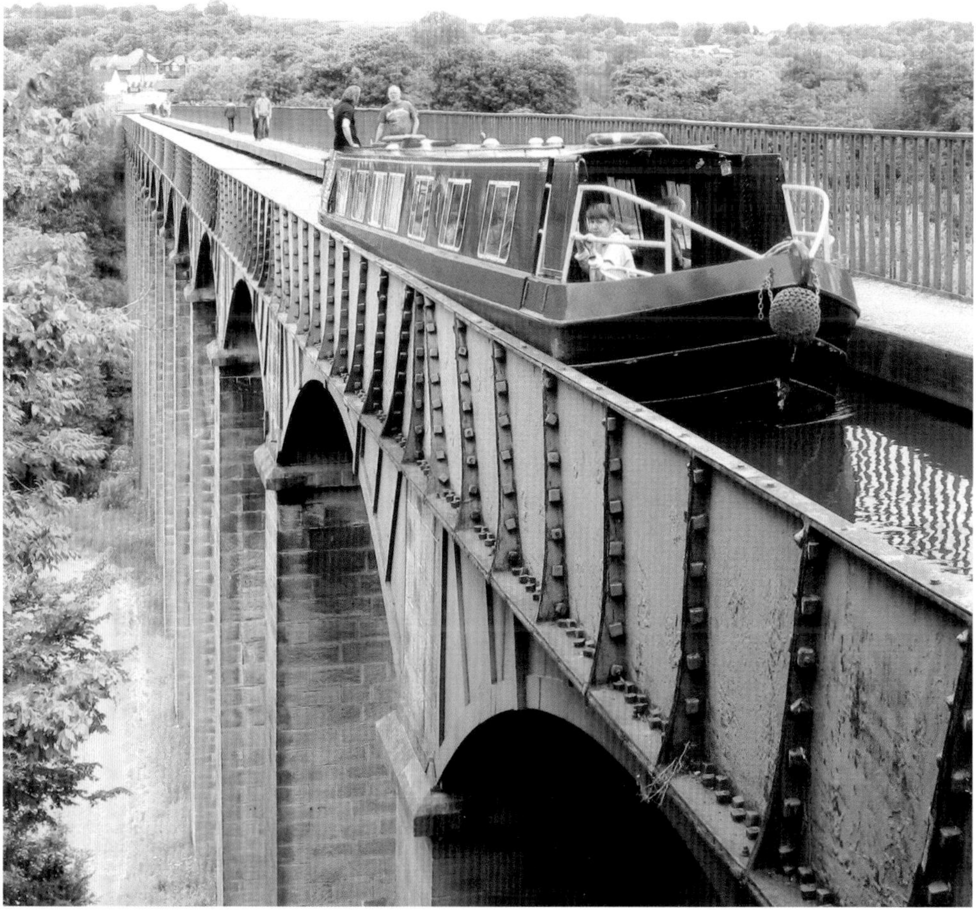

Pontcysyllte (Wikipedia/Arpingstone).

The Pontcysyllte Aqueduct, dramatic climax of an 11-mile stretch of the Llangollen Canal in north-east Wales, was completed by Thomas Telford in 1805. The elegant structure, sky-high above the River Dee, a masterpiece by one of Britain's most innovative and courageous engineers, was elevated to UNESCO World Heritage status in 2009. Our own small adventure still ranks as the high point of our 25-year love affair with narrowboats on Britain's inland waterways. We will not forget the day we launched out across Telford's 'stream in the sky'.

But what exactly is meant by 'aqueduct'? It comes from the Latin *aqua* (water) and *ducere* (to lead), and is often used to describe a bridge designed for water supply, navigation, or both – without necessarily including the adjoining canal or waterway. More generally it refers to any artificial channel, conduit or pipe that carries water from one place to another, below ground, at ground level or raised on a structure.

Our story begins with a brief account of five ancient aqueducts that lead, step by step, from subterranean tunnels via ground-level channels to elevated bridges. All enjoy World Heritage status, and illustrate the extraordinary efforts made by human societies to bring reliable supplies of that most essential commodity of all – fresh water – to their towns and cities. The final example is a spectacular Roman masterpiece that was well known to British canal engineers in the 18th and 19th centuries, including Thomas Telford; and although its purpose was water supply rather than navigation, it shared with Pontcysyllte the challenge of supporting huge quantities of water at dizzying heights above a valley.

The Aqueduct of Eupalinos in Samos, Greece (Wikipedia/Moumouza).

Although we tend to associate historic aqueducts with Ancient Rome, many were built centuries earlier in countries such as India, Persia, Iraq, Egypt and Greece, allowing societies to grow and flourish in some of the most arid regions of the world. To take just two examples: in the 7th century BC the Assyrians built an extensive limestone aqueduct, including a high section across a valley, to bring water to their capital city of Nineveh, located in modern-day Iraq; and in the 6th century BC a Greek named Eupalinos excavated a tunnel under Mount Kastro on the island of Samos to carry water to the nearby city of Pythagoreion. It remained in service for 1,000 years, then lay dormant. Today it is a major tourist attraction.

The Greek historian Herodotus (484–425 BC) was fascinated by the aqueduct that Eupalinos had built on Samos:

> first a passage beginning from below and open at both ends, dug through a mountain not less than a hundred and fifty orguia in height; the length of the passage is seven stadia and the height and breadth each eight feet, and throughout the whole of it another passage has been dug twenty cubits in depth and three feet in breadth, through which the water is conducted and comes by pipes to the city, brought from an abundant spring.

Eupalinos gave his tunnel a gentle downward slope along its 4,000-foot (1,040 metres) length, allowing the water to flow by gravity. Most remarkably, he started excavation from both ends and used geometric principles to ensure that his two teams of miners met in the middle. Specifically, he realised that two straight parallel sections would fail to meet if their horizontal offset exceeded the tunnel width. So as the teams got closer, he ordered them to change direction, one slightly to the left and the other to the right, creating a dogleg and guaranteeing an intersection. Vertically, he was more confident of accurate alignment; even so – better safe than sorry – he increased the height of both sections as the teams converged, raising the roof of one and lowering the floor of the other. Eupalinos's insights had to wait several centuries before being formalised by Euclid in his famous geometry book, the *Elements*.

For our next example we travel to Jordan, where ground-level aqueducts formed part of an elaborate water supply system in the ancient city of Petra. Its stunning architecture, fashioned out of solid rose-tinted rock, was created from the 4th century BC onwards by the Nabateans, a nomadic Arab people who inhabited northern Arabia and the Levant.

Petra's location close to regional trade routes made it increasingly wealthy, but this was only possible because the sparse local rainfall and occasional flash floods were efficiently captured by the Nabateans and stored in dams and cisterns. Then, on flattish ground near the city centre, gently sloping open aqueducts distributed water to public buildings, homes and gardens.

In the 1st century AD Petra and its territory were annexed by the Roman Empire, marking the end of the Nabatean dynasty. Initially the city flourished under Roman rule, to reach the height of its splendour. But later it entered a long period of decline as Palmyra

Remains of the ancient aqueduct system in the city of Petra (Wikipedia/Etan J. Tal).

(in modern-day Syria) grew in importance. Petra remained unknown in the West until 1812, when its buildings and aqueducts were discovered by a Swiss explorer. UNESCO awarded the city World Heritage status in 1985, describing it as 'one of the most precious properties of man's cultural heritage'.

For our third example we travel to Hampi in the state of Karnataka in south-west India, where an ancient bridge-like aqueduct has survived the ravages of time in remarkably good order. The village and temple town of Hampi is located within the ruins of the city of Vijayanagara, the former capital of an empire which ruled over a vast area of southern India between 1343 and 1565 AD. In its prime Vijayanagara was one of the richest and largest cities in the world, with about half a million inhabitants. It was completely dependent on aqueducts for crop irrigation and water supply to temples and other buildings, including a royal palace that pampered its owners with luxurious baths. The Hampi Aqueduct formed part of a remarkable system that included canals, sluices, storage tanks, gravity-fed channels and siphons.

The ruins of Hampi were explored by Colonel Colin Mackenzie in 1800. Born on the Isle of Lewis in Scotland, he more than lived up to family expectations by becoming Surveyor-General of India and a famous archivist of India's antiquities. Today Hampi and Vijayanagara are classed as a UNESCO World Heritage Site.

Of all the early civilisations that built aqueducts, the best known in Europe is undoubtedly that of the Ancient Romans. Over several centuries they built nine major aqueducts to serve their capital city, with a total length of 260 miles (420 km). Water was obtained from distant streams and springs and, on arrival in Rome, delivered to temples and civic buildings, public baths, latrines and access points for collection by private citizens. The longest aqueduct was the Marcia, built between 144 and 140 BC with a drop of 850 feet (260 metres) over its length of 57 miles (91 km) – an average slope of 1 in 350.

Roman aqueducts followed a gentle downhill course at ground level wherever possible. Contrary to popular belief, bridges were used sparingly, because they were extremely costly to build and maintain. Frontinus, Rome's water commissioner at the end of the 1st century AD, left a famous account of its water system, including his beloved aqueducts:

> With such an array of indispensable structures carrying so many waters, compare if you will, the idle Pyramids or the useless though famous works of the Greek.

Frontinus was clearly a practical man without much feeling for art or architecture. Of greater concern to him was the loss of precious water on its journey to Rome:

Hampi Aqueduct, India (Wikipedia/Somsubhra Sharangi).

The Aqueduct of Segovia (Wikipedia/Bernard Gagnon).

> A large number of landed proprietors, past whose fields the aqueducts run, tap the conduits; whence it comes that the public water courses are actually brought to a standstill by private citizens, who water their gardens.

The Romans went on to build aqueducts in all parts of their empire, from Western Europe to North Africa and the Middle East, setting standards of engineering that would not be surpassed for 1,000 years. Stone bridges with multiple arches were among the most impressive features, and of those that survive a superb example is to be found in central Spain.

The city of Segovia's aqueduct was built to serve a civilian population that followed in the footsteps of a victorious army, described by Frontinus as 'the most solemn testimony of the Empire'. It transported water from the Rio Frio, situated in mountains about 11 miles (17 km) distant, and it still dominates the city centre, with two tiers of splendidly preserved arches. The date of construction was long a mystery but recent evidence points to about 110 AD. The water conduit at the top of the structure – U-shaped, and with a cross-section of about 2.7 square feet (0.25 square metres) – provided Segovia with a vital water supply well into the 19th century.

The bridge section of the aqueduct has a total length of about 900 yards (813 metres) and consists of four straight segments which change direction as they approach the city centre. No less than 167 tiered arches, built of granite blocks without mortar, reach skywards, with heights up 93 feet (28.5 metres). They were constructed following principles laid down by Vitruvius, a Roman architect and military engineer, in a 1st-century treatise on architecture and building technology dedicated to his patron, the Emperor Caesar Augustus.

Buildings in Segovia's Old Town, huddled together on a rocky bluff, form a prime example of the coexistence of different cultural communities – the Moors, Christians and Jews who lived and worked together for centuries. The layout of neighbourhoods, streets, and houses reflects their hierarchies and social organisation, and evidence of peaceful coexistence is strengthened by the large number of ancient buildings and monuments that have survived. About half a square mile (134 hectares) of this remarkable city was awarded UNESCO World Heritage status in 1985, with the Roman Aqueduct as its dominant feature.

Our final example of an ancient aqueduct made a big impression on me as a teenager on holiday in the south of France. I was amazed by the massive scale and stunning architecture of the Pont du Gard, but understood little of its history or purpose. Actually, it is another classic example of the Roman bridge aqueduct, and it deserves a good slice of our attention, because it was well known to Thomas Telford.

The Pont du Gard is about 15 miles (24 km) from the famous city of Avignon and dates from about 60 AD. Added to UNESCO's list of World Heritage Sites in 1985, it is the highest

The Pont du Gard (photo: Jessica Armstrong).

Route of the Nîmes Aqueduct between Fontaine d'Eure near Uzès and the city of Nîmes (Wikipedia/Clem Rutter); and the Pont du Gard's water conduit (Wikipedia/Édouard Bergé).

of all surviving Roman bridge aqueducts and one of the best preserved. It formed part of the 30-mile (48 km) Nîmes Aqueduct that delivered water from a spring near the town of Uzès to the Roman colony of Nemausus, the modern city of Nîmes. Fed by gravity, it had an astonishingly small drop of 56 feet (17 metres) over its entire length, giving an average slope of just 1 in 2,800. The terrain between its extremities was so uneven that the Romans chose a winding route to circumvent the worst obstacles; but when they came to the deep gorge of the Gardon river, there was no alternative but to preserve precious height with a bridge. (More bridges were needed a few miles downstream, but nothing on so majestic a scale.)

The bridge's three tiers of limestone arches tower 160 feet (48.8 metres) above the river. The conduit, 4 feet (1.2 metres) wide by 6 feet (1.8 metres) high, descends a mere 1 inch (2.5 cm) over its length of 902 feet (275 metres) and stands testament to the great precision achieved by Roman engineers using the most basic surveying instruments. It carried up to 44 million gallons (200,000 cubic metres) of water a day to the buildings, fountains, public baths and homes of Nîmes. If we divide the daily flow by the cross-sectional area (say 2 square metres) we get an average speed of about 4 km per day, or 2.5 mph, along the conduit.

As the mighty Roman Empire teetered towards collapse, lack of maintenance caused clogging by debris and eventually choked off the flow. But the structure remained largely intact due to its use as a toll bridge. Local landowners and bishops took responsibility for upkeep in exchange for charges levied on travellers, goods and livestock. Over the centuries there was some damage and looting of stones, but the bridge was repaired a number of times from the 17th century onwards, and the bottom tier was increased in width to allow a proper roadway to be added.

By the mid-18th century the Pont du Gard was being lauded again – more as a sublime expression of classical architecture than as a water carrier. This was the age of the Grand Tour, the traditional journey through France and Italy by wealthy young men from north-western Europe (and especially Britain) in search of art, architecture and the roots of Western civilisation. We may imagine them calling in at the Pont du Gard as they travelled on their educational rites of passage between historic cities, including Avignon and Nîmes, accompanied by coachmen, valets, perhaps cooks, and certainly scholarly guides. They set a trend, and the much-admired aqueduct remains a tourist attraction to this day.

Grand Tourists returning home helped stimulate the Neoclassical movement in art and architecture that flourished in western Europe, particularly England, between about 1750 and 1850. It influenced such famous London buildings as Buckingham Palace, the Bank of England and Somerset House, and affected garden design by introducing classical features into the parks and gardens of grand country houses. A fine example of the English 'landscape garden' is Stourhead in Wiltshire, created by banker Henry Hoare, who travelled to France and Italy on the Grand Tour and returned home the proud owner of a pastoral landscape by French artist Claude Lorrain. Smitten by classical design, Hoare ordered the damming of a stream and created a garden with lake, trees and neoclassical architecture.

Neoclassical design: Stourhead in Wiltshire, England (Geograph/Colin Smith).

The Neoclassical movement also influenced the civil engineering profession which in the second half of the 18th century was at the point of take-off. The first formal meeting of British civil engineers took place in a London tavern (no surprise there) in March 1771 with John Smeaton, designer of bridges, canals and lighthouses, as its leading light. However, in time his Society of Civil Engineers morphed into an exclusive dining club rather than a professional body, and a younger group of men began to press for something more useful. The Institution of Civil Engineers was founded in 1818, and Thomas Telford, already world-famous, agreed to be its first president. He was determined to make it a fount of engineering knowledge and achievement that would promote civil engineering as a profession. But he was also concerned with aesthetics: like other famous British engineers, including John Rennie and Isambard Kingdom Brunel, he was well aware of Neoclassical architecture and the crisp elegance it could bring to bridges and other massive structures.

Thomas Telford's earliest English bridge was completed when he was 35 years old. The Montford Bridge crosses England's longest river, the Severn, close to the town of Shrewsbury in Shropshire, and was built to carry his London to Holyhead road. It has three arches of local sandstone, one with a span of 58 feet and two of 55 feet, and looks splendid from the riverbank. Telford remained proud of it for the rest of his life, and as an old man noted that 'it has proved a substantial edifice, having been completed upwards of 40 years, and remaining quite perfect'.

The Montford Bridge was widened in 1963 and although most traffic is now diverted onto the Shrewsbury bypass, it still carries a stream of vehicles that would have been unimaginable to its designer over two centuries ago.

When we come to canal aqueducts, the earliest English examples used plain arch bridges and were built in the 1760s by the great engineer James Brindley (1716–1772). But by the early 1790s 'Canal Mania' was at its height, and some professional engineers were keen to

Telford's Montford Bridge over the River Severn (Geograph/Kevin Skidmore).

The Dundas Aqueduct over the River Avon near Bath (Geograph/John Winder).

embellish their designs with classical features. A fine example is the Dundas Aqueduct, which carries the Kennet and Avon Canal over the River Avon about 9 miles (15 km) from Bath, England's most neoclassical city. Built by John Rennie, the aqueduct was completed in 1810 – five years after Pontcysyllte – with classical decorations that suggest an elegant permanence. Rennie clearly wished to proclaim his massive aqueduct an artistic addition to the landscape, as well as a functional structure. Unfortunately, however, this is not evident to narrowboat crews and towpath walkers unless they, like visitors to the Montford Bridge, climb down to the river bank for a full appreciation.

How does all this relate to Pontcysyllte, and what possibilities suggested themselves to Thomas Telford as he approached the enormous task of carrying the Llangollen Canal over the River Dee? Would he attempt an English Pont du Gard with tiered stone arches; or would he descend with a flight of locks, cross the river at low level, and climb up with more locks on the far side? Perhaps he would strike out in an entirely new direction. The short answer to a long question is that he had become increasingly fascinated by two words: *cast iron*.

The first substantial cast-iron arch bridge in the world was opened for business on New Year's Day in 1781 – and the 24-year old Telford knew all about it. Located in the village of Ironbridge, near Coalbrookdale in the county of Shropshire, the bridge now forms part of the Ironbridge Gorge UNESCO World Heritage site, which includes mines, factories, workers' housing, and transport infrastructure created in the earliest decades of Britain's

Crossing the Dundas Aqueduct (Geograph/Stephen McKay).

The iron bridge in Ironbridge (Geograph/David Dixon).

Industrial Revolution. The gorge attracts over half a million visitors a year, and although the bridge no longer carries horses and carts over the Severn, it is very much open to pedestrians and sightseers.

It is hard to imagine the frenzy of activity accompanied by noise, fire and smoke that disturbed the rural peace of this part of Shropshire in the late 18th century. Abraham Darby I had first smelted local iron ore with coke made from Coalbrookdale coal in 1709. By 1776 his grandson, Abraham Darby III, a local ironmaster, was being commissioned to cast and build a bridge with a 100-foot span that would allow free flow to the river and the many boats that traded along it. The result, a structure of wonderful delicacy allied to strength, soon became one of the most famous bridges in the world.

From this moment on the possibilities offered by cast iron fascinated Thomas Telford. Sixteen years later he chanced his arm by building the world's first substantial iron aqueduct on the Shrewsbury Canal at Longdon-on-Tern, replacing a conventional aqueduct that had been swept away by floods two years earlier. The canal has long been abandoned, but the 187-foot (57 m) aqueduct remains in surprisingly good condition, curiously isolated among fields astride the River Tern; hardly a 'stream in the sky' – but, like the Hampi Aqueduct, definitely above ground level.

Shortly after completing Longdon-on-Tern, Telford, one of the world's most daring and creative engineers, eschewed the tiered stone aqueducts of his Roman predecessors

Telford's cast-iron aqueduct at Longdon-on-Tern (Geograph/John Horton).

Pontcysyllte, Telford's 'stream in the sky' (Geograph/Steve Daniels).

and dared to use the new material at Pontcysyllte. His very own 'stream in the sky' still stands elegantly over the River Dee more than 200 years later, and has recently confirmed the tendency of historic aqueducts to attract international recognition by being awarded UNESCO World Heritage status.

But I am in danger of putting the narrowboat before the horse, because Telford was 35 when he built the Montford Bridge, and 48 – advanced middle age by the standards of the time – when he completed Pontcysyllte. To understand why this remarkable man became so heavily involved in transport infrastructure, we must cast back to the 18th century and discover how raw materials and finished products were moved around the country as Britain's industrial revolution gathered pace.

2 Troublesome Travel

The English canal system is much loved by boaters, anglers, walkers and cyclists. Of course there is occasional criticism: locks hard to operate; water in short supply on summit levels during the summer months; moorings blocked by selfish boaters overstaying their welcome. But such has been the remarkable restoration work carried out in the last 50 years, much of it by volunteers for the benefit of leisure users, that few of us have the heart to complain.

By contrast it is almost a national pastime to grumble about the state of the roads. There is a more or less annual clamour for thousands of potholes to be filled in – not patched up with a quick dollop of some ineffective material, but filled in *properly*. Motorists worry about damage to their suspensions, cyclists about being catapulted over their handlebars. It is all very troublesome.

Canals and roads were also topics for intense discussion in the 1760s – the years of Thomas Telford's childhood. But it was mostly talk, not action, and the shocking state of the road network meant that packhorses were still the main means of bulk transport overland. Unfortunately, beasts of burden were no answer to manufacturers wanting to send steel bars from Sheffield to Bristol, or delicate china teapots from Stoke-on-Trent to London. Arteries for Britain's burgeoning industrial revolution were urgently needed, and together with the main rivers, these would be provided by canals and much better roads.

In 1760 Francis Egerton, Duke of Bridgewater, exasperated by the slowness, unreliability and inefficiency of the packhorse trains that carried coal eight miles from his mines in Worsley to the expanding town of Manchester, decided to commission a canal. The flat terrain required no locks, but the River Irwell was in the way and had to be crossed. The duke, a Grand Tour enthusiast, was well aware of the navigable aqueducts on the Canal du Midi in France, but the very idea of a bridge carrying a canal over a river seemed ridiculous to many of his English contemporaries, including certain engineers. As one brought in to review his plans commented: 'I have often heard of castles in the air, but never before saw where one was to be erected.' Yet the duke pressed ahead with the Bridgewater Canal, and appointed 44-year-old James Brindley as Engineer.

Born in 1716 into a family of yeoman farmers in the Peak District of Derbyshire, Brindley had been educated at home by his mother before becoming a millwright, a wheelwright, and eventually a talented repairer and even designer of new types of machinery. His first task for the Duke of Bridgewater was to design and build a bridge aqueduct over the River Irwell near the village of Barton.

In January 1760 Brindley was dispatched to London to champion his Barton Aqueduct before a parliamentary committee. The politicians were puzzled by his intention to use clay for a watertight seal, and irritated by his reluctance to commit plans to paper. Clearly bent on enjoying himself, Brindley had a mass of clay brought into the committee room, formed it into a trough and demonstrated its leakproof qualities by 'puddling' it with water. When asked to produce a drawing of the aqueduct, he purchased a large round Cheshire cheese, cut it into semicircles, inverted them, and placed a board on top to represent the canal. 'Here', he told the committee, 'is my model.'

The necessary Act of Parliament, passed in March 1760, was specific about the form of the aqueduct: it must have three arches at least as wide and high as those of a nearby road bridge to prevent restricting river traffic on the Irwell. Brindley came up with a design about 600 feet long and 36 feet wide, an enormous structure for its time. It was completed in just 11 months; but on the first day water was admitted it began to sag. Poor James Brindley, overcome with anxiety, retired to his bed at the Bishop Blaize tavern in nearby Stretford and left a colleague to address the problem with a partial rebuild. The aqueduct was re-puddled and allowed to stand until the following spring, by which time the mortar had set and the danger of collapse had passed.

Crowds came from all over the country to view a wonder of the age, amazed to see a single horse hauling linked barges along Brindley's aqueduct while a team of horses struggled to haul a single barge against the flow of the Irwell 40 feet below. Just as importantly for the Duke of Bridgewater and his customers, the new-fangled 'navigation' cut transport costs to such an extent that the price of coal in Manchester fell by a half.

The fate of Brindley's famous structure was eventually sealed by the Manchester Ship Canal Act of 1885, which approved a 36-mile navigable waterway for ocean-going

The Barton Swing Aqueduct across the Manchester Ship Canal (Geograph/David Dixon).

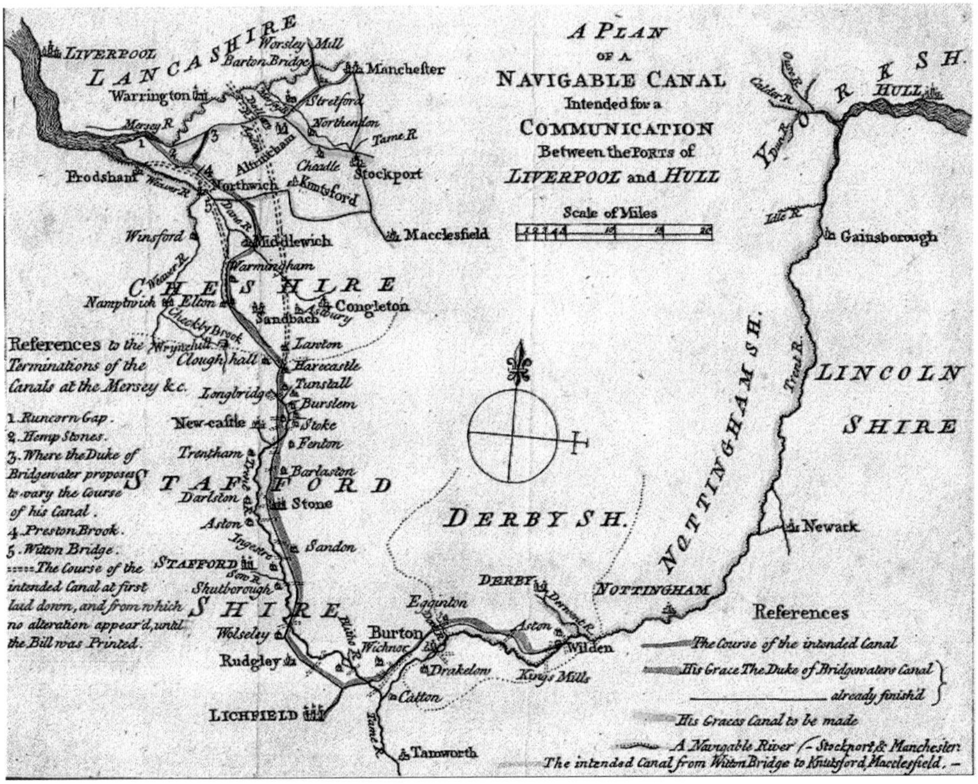

A historic map showing the proposed route of the Trent and Mersey Canal. The ports of Hull and Liverpool are shown at top right and top left respectively (Wikipedia).

vessels from the estuary of the River Mersey into Manchester. It incorporated a section of the River Irwell through Barton, but the arches of Brindley's aqueduct were too small for large ships to pass beneath, so after 132 years' service it was demolished with the help of dynamite.

Its replacement is worth a slight digression because it was, and at the time of writing remains, the world's only *swing* aqueduct. Designed to rotate on a massive pivot set into a small purpose-built island, the Barton Swing Aqueduct was the brainchild of Sir Edward Leader Williams, Engineer to the Manchester Ship Canal Company. The first barge crossed his remarkable structure in August 1893. Normally kept open to allow large vessels along the ship canal unhindered, it still gives temporary precedence to narrowboats and other craft on the Bridgewater Canal by swinging its 330-foot, 1,430-ton trough through 90 degrees. Stop gates retain around 800 tons of precious water in the trough, and seal adjacent stretches of canal. We may imagine Sir Edward in his design office, poring over plans for a huge iron trough and proud to be following in the footsteps of Thomas Telford.

James Brindley's growing fame put him in great demand, and he was soon appointed Engineer to the 93-mile Trent and Mersey Canal linking the River Trent near Nottingham to the Bridgewater Canal and the River Mersey – a project to which he would devote six years of unremitting attention before his untimely death in 1772.

The Trent empties into the North Sea and the Mersey into the Irish Sea, so proponents of the scheme, considered hare-brained by their adversaries, championed it as an east–west link between the sea ports of Hull and Liverpool. Begun in 1766 and opened for business in 1777, it has no aqueducts but includes 76 locks and 4 tunnels. At first sight its route seems bizarre. Liverpool is roughly due west of Hull, but the complete waterway embarks on a giant curve to the south, making full use of the navigable River Trent and skirting around Brindley's hilly home country, the Peak District of Derbyshire. It comes within 25 miles of Birmingham before striking north through Staffordshire and on towards the Mersey. The route was designed to minimise the number of locks – a predilection that was to become one of Brindley's trademarks.

Another consideration, commercial rather than geographical, supported the circuitous route. The potteries around Stoke on Trent were in desperate need of something better than packhorses to export their fragile wares, so Josiah Wedgwood, the most famous potter of them all, became a wholehearted advocate of a connection to the new canal. Indeed so much so that he cut the first ceremonial sod in 1766, leaving Brindley to wheel it away in a barrow. They became good friends, especially when Brindley and his surprisingly young wife, Anne, took up residence in Turnhurst, a fine house near the planned summit of the canal and Wedgwood's soon-to-be factory at Etruria.

Brindley was new to locks, so he decided to experiment with a model in the grounds of Turnhurst. The design he came up with was unusually narrow, able to take a barge up to

Long and narrow: Brindley's Lock no. 41, just north of the Harecastle Tunnel in Staffordshire (Geograph/Star Watcher).

A 72-foot narrowboat fills Stoke Top Lock on the Trent and Mersey Canal (Geograph/Andy Beecroft).

The northern entrances to Thomas Telford's Harecastle Tunnel of 1827 and, in the background, James Brindley's tunnel of 1777, now abandoned (Wikipedia/Akke Monasso).

72 feet long but only 7 feet wide. His decision was to have enormous consequences for the English canal system and, some say, cast a long shadow over it.

He was also new to tunnels. Of the four needed, the Harecastle in Staffordshire, one of the longest canal tunnels ever built in England, was by far the most problematic. Beset by unexpected changes of rock type, flooding, and lack of ventilation, its seven-year construction outlasted Brindley by five years. He gave it no towpath, expecting boatmen to 'leg' their way through by lying on the top of the boat and pushing on the slimy sides or roof of the tunnel with their feet. The 2,880-yard passage took up to three hours, and they would emerge at the far end completely exhausted and 'as wet from perspiration as if they had been dragged through the canal itself'. Meanwhile, the lucky horses were escorted over Harecastle Hill.

Growing traffic and the painfully slow process of legging soon made the Harecastle Tunnel a serious bottleneck. Boats would pile up at one end, waiting for others to complete their passage in the opposite direction, and there were many furious battles between claimants for the 'first turn through'. So in 1824 it was decided to build a wider, parallel, tunnel with a towpath for the horses. Thanks to advances in civil engineering and the talents of its designer – none other than Thomas Telford – it was completed in three years. The two tunnels were restricted to one-way traffic and used in conjunction, but following a partial collapse of Brindley's tunnel in 1914 Telford's took over completely. Nowadays narrowboats complete the passage in about 40 minutes, their crews protected from diesel fumes by powerful forced ventilation.

In the last few years of his life Brindley planned, and partly completed, more than 300 miles of canals, including the Staffordshire and Worcestershire, the Coventry, and the Oxford. He favoured routes that followed the contours of the land, minimising locks and tunnels. However this came at the cost of long meanders through the countryside. If you have navigated the Oxford Canal you will know exactly what this means: on one particularly circuitous section south of Napton Junction the canal winds apparently aimlessly for 11 miles between two points just 5 miles apart.

Brindley's early successes proved the viability of horse-drawn canal boats, and industrialists in other parts of the country were soon clamouring for a piece of the action. Within a few years an embryonic national network was in place, especially in the Midlands and the north of England, where the navigable rivers badly needed interconnecting. Initially the traffic was mainly inland; but as the network reached coastal port cities, including London, Liverpool and Bristol, more cargo was exchanged with seagoing ships for import and export. As the 18th century drew to a close almost all Britain's finest civil engineers were involved. High reputations and a great deal of money were to be made out of successful canals.

However there was also frantic speculation. People bought shares in canal companies simply to sell them on for immediate profit, regardless of whether a canal was viable, or indeed even built. Huge sums were invested during the years of canal mania, and many schemes came to nothing. Yet in time the network expanded to more than 3,000 miles. Unfortunately, however, in the long run Brindley's narrow locks, adopted as the standard for much of the system, and limiting the quantity of cargo carried by a single barge to about 30 tons, made the network less and less competitive for bulk transport when the railway revolution gathered pace in the 1840s.

Judged by today's standards, we may wonder why the ability to horse-haul 30 tons of coal, sand, china clay or manufactured goods at average speeds of about 2 miles per hour was considered such a huge improvement on existing modes of transport. Surely wheeled vehicles on improved roads would be a faster and better solution; and where new roads were needed, surely they could be built more cheaply than canals by avoiding the huge expense of locks, tunnels, embankments, and the occasional aqueduct?

Not necessarily. Most of the existing road network was in a shocking state of disrepair, and useless for wheeled vehicles. Effective and affordable ways of putting it right were little understood, and so packhorses were still the main means of bulk transport overland. This may be hard to comprehend – but not if we explore the fascinating account of 'troublesome travel' left by a remarkable Victorian author.

We will do well to take notice of Samuel Smiles (1812–1904), not only for his review of travel arrangements in the 18th and 19th centuries, but for his knowledge of the life and work of Thomas Telford. Among engineering aficionados Smiles is best known for his five-volume series, *Lives of the Engineers*, published in London in 1862, a brilliant multiple biography covering a galaxy of engineering talent from James Brindley via Thomas Telford to George and Robert Stephenson of locomotive fame. Less well known is that the third volume in the series, *History of Roads*, was expanded five years later with *An Account of Early Roads and Modes of Travelling in Britain*. For an explanation of the conditions

A plaque to Samuel Smiles in London (Wikipedia/Spudgun67).

that gave rise to canal mania, it is something of an open sesame – or perhaps we should say a Pandora's box.

By the time Smiles had completed his *Lives of the Engineers* at the age of 50 he had already achieved far more than most people manage in a lifetime. Born in East Lothian, Scotland, one of eleven children, he became a man of many parts. After studying medicine at Edinburgh University he developed an unlikely passion for political journalism, especially as it related to social and economic conditions in Victorian Britain. By the age of 30 he was Editor of the Leeds Times and Secretary to the Leeds Parliamentary Reform Association, a Chartist organisation with an agenda that for its time was little short of revolutionary. Three years later he branched out again, to become Secretary of the newly formed Leeds & Thirsk Railway, and subsequently of the South Eastern Railway. By this time his passion for parliamentary reform was spent, displaced by a growing conviction that individual self-improvement was a surer remedy for society's many ills. And when in 1859 his book *Self Help* was published he became, at the age of 47, a national celebrity.

The origins of *Self Help* lay in a speech given to a Mutual Improvement Society claiming that poverty was caused largely by irresponsible personal behaviour. Social progress would come from new attitudes rather than new laws. However, lest this be thought unduly harsh and against his Chartist sympathies, he also attacked rampant materialism and the injustices of *laissez-faire* capitalism. The book has been called 'the bible of mid-Victorian liberalism'.

By the time *Lives of the Engineers* appeared three years later, it seems that his experience as secretary of two new railway companies had stimulated an interest in the engines of the industrial revolution and, more importantly from our point of view, in the remarkable characters who made it possible. An author veering towards the primacy of self-improvement over government action for the progress of society would have found plenty to applaud in the great engineers who, in most cases, reached the top of their profession without any privileges of wealth or family background – among them James Brindley, and especially Thomas Telford.

Samuel Smiles' writing on 'troublesome travel' makes fascinating reading. His style, somewhat reminiscent of Charles Dickens, propels us back to Victorian times in ways that a modern writer would struggle to achieve – slightly quaint, but thoroughly engaging, and redolent of a very different age:

> Freedom itself cannot exist without free communication – every limitation of movement on the part of the members of society amounting to a positive abridgment of their personal liberty. Hence roads, canals, and railways, by providing the greatest possible facilities for locomotion and information, are essential for the freedom of all classes, of the poorest as well as the richest.

His historical notes on English roads go right back to the Romans, but we may skip to the time of the first Queen Elizabeth:

> The great Western road into London was especially bad, and about Knightsbridge, in winter, the traveller had to wade through deep mud. Wyatt's men entered the city by this approach in the rebellion of 1554, and were called "draggle tails" because of their wretched plight … At a greater distance from the metropolis, the roads were still worse. They were in many cases but rude tracks across heaths and commons, as furrowed with deep ruts as ploughed fields; and in winter to pass along one of them was like travelling in a ditch.

The situation was hardly better by the mid-18th century, even close to rapidly expanding Birmingham, 'now the centre of an immense traffic'. Roads had been worn down by generation after generation of human feet and horse hooves, scoured by the rains and often gouged so deep that a rider on horseback would disappear below the level of the ground on either side. The roads of Sussex, in particular, enjoyed an 'infamous notoriety', making it harder for old people to get to church in winter than it was across the sodden Lincolnshire Fens. One observer claimed that Sussex girls were exceptionally long-limbed because the tenacity of mud in the county tended 'to stretch the muscle and lengthen the bone'. Closer to London, just a few years before James Brindley presented his plans for Barton Aqueduct to Parliament, a resident of Kensington complained that:

> the road between this place and London is grown so infamously bad that we live here in the same solitude as we would do if cast on a rock in the middle of the ocean; and all the Londoners tell us that there is between them and us an impassable gulf of mud.

To which Samuel Smiles adds:

> Nor was the mud any respecter of persons; for we are informed that the carriage of Queen Caroline could not, in bad weather, be dragged from St. James's Palace to Kensington in less than two hours, and occasionally the royal coach stuck fast in a rut, or was even capsized in the mud … the streets in the towns and the roads in the country were

> alike rude and wretched, indicating a degree of social stagnation and discomfort which it is now difficult to estimate, and almost impossible to describe.

But how about wheeled vehicles other than royal carriages?

Simple unsprung wagons hauled by teams of horses had been in use for centuries for transporting goods and, to a lesser extent, people; but passengers found the mode of travel:

> so tedious, by reason they must take waggon very early and come very late to their innes, that none but women and people of inferior condition travel in this sort … The waggons made only from ten to fifteen miles in a long summer's day; that is, supposing them not to have broken down by pitching over the boulders laid along the road, or stuck fast in a quagmire, when they had to wait for the arrival of the next team of horses to help drag them out. The waggon, however, continued to be adopted as a popular mode of travelling until late in the eighteenth century.

And how about stagecoaches?

We have become used to films showing elegant Jane Austen characters inside a comfortable stagecoach, with driver and less affluent individuals be-rugged on the outside, the whole contraption hauled by spirited horses at a respectable 10 or 12 miles per hour. But this was the Regency period, the golden age of stagecoaching between about 1810 and 1830; when James Brindley started on the Bridgewater Canal in 1760, things were a great deal less comfortable – indeed, they had barely improved in the 100 years before that:

> The introduction of stage-coaches about the middle of the seventeenth century formed a new era in the history of travelling by road. At first they were only a better sort of waggon, and confined to the more practicable highways near London. Their pace did not exceed four miles an hour, and the jolting of the unfortunate passengers conveyed in them must have been very hard to bear. It used to be said of their drivers that they were "seldom sober, never civil, and always late".

Other discomforts could be avoided by using the 'more noble method of travelling by horseback'. As one commentator exclaimed:

> What advantage is it to men's health to be called out of their beds into these coaches, an hour before day in the morning; to be hurried in them from place to place, till one hour, two, or three within night; insomuch that, after sitting all day in the summer-time stifled with heat and choked with dust, or in the winter-time starving and freezing

Dream and reality: an elegant Royal Mail coach travelling at speed, and a heavy wagon struggling along the heavily rutted Great North Road before turnpiking was introduced (Wikipedia).

> with cold or choked with filthy fogs, they are often brought into their inns by torchlight, when it is too late to sit up to get a supper … What addition is it to men's health or business to ride all day with strangers, oftentimes sick, ancient, diseased persons, or young children crying … poisoned with their nasty scents and crippled by the crowd of boxes and bundles … to be affronted by the rudeness of a surly, dogged, cursing, ill-natured coachman?

To be fair, there were some improvements in road travel from the mid-18th century onwards: better-designed coaches with leaf-spring suspension and steel-rimmed wheels; better maintenance of the principal routes by the rapidly expanding Turnpike Trusts, set up by acts of Parliament to collect tolls. But even so, comfortable travel by road was for most citizens a distant dream. And the final, inescapable, truth is that when it came to transporting large quantities of goods and materials over difficult terrain, neither heavy wagons condemned to battle along rutted roads that half-buried their wheels, nor coaches intended for people and their luggage, could compete with packhorses. As the industrial revolution took off, packhorse convoys were still, as they had been for centuries, England's main means of bulk transport.

Poor dumb creatures – why were the Duke of Bridgewater and other new industrialists so fed up with the service they provided? The main reason is also the obvious one: the small weight of goods a packhorse could carry – about 240 pounds, distributed evenly between two side-panniers strapped across its back. True, packhorses could be arranged in long convoys, typically 12 to 20 and occasionally up to 40, and coaxed or bullied as much as 25 miles a day along tracks and roads that were impossible for wheeled vehicles. But compare this with a 72 × 7-foot barge that fitted snugly into one of James Brindley's new-fangled locks: although it travelled at roughly the same speed, its load would be about

A packhorse convoy (Wikipedia).

30 tons or 67,000 pounds – equivalent to 280 packhorse loads – transportable by a single horse and a crew of just two.

Packhorses had other disadvantages. The small stocky animals, often named Galloways after the Scottish district where they were first bred, needed a lot of feeding and human attention; many were lamed by their arduous duties; and a train was often impeded by dreadful conditions underfoot or the difficulty of passing others travelling in the opposite direction. The animals often wore loops or collars of bells to warn of the convoy's approach:

> in many parts of the path there was not room for two loaded horses to pass each other, and quarrels and fights between the drivers of the packhorse trains were frequent as to which of the meeting convoys was to pass down into the dirt and allow the other to pass along the bridleway.

According to Samuel Smiles, conditions in the 1760s were hardly better than 100 years before and led to innumerable quarrels. Injuries were common, deaths occasional. Some packhorse men, 'rude Russian-like rake-shames', simply got rid of approaching horseback riders with booze-fuelled threats of violence.

Much of England's early industrialisation was concentrated in the north. By the mid-18th century huge numbers of packhorse convoys were crossing the Pennines loaded with coal, limestone, salt, fleeces and cloth. Up to 1,000 packhorses a day passed through the small Lancashire town of Clitheroe. The importance of major routes was reflected in popular jingles and rhymes that served as *aide-memoires* for the packhorse men, and in the growing number of Packhorse Inns, many of which have survived into our own times.

The ancient Packhorse Inn at Low Leighton, looking towards Derbyshire's Peak District (Geograph/Bob Harvey).

In some areas, especially those with harsh terrain, the use of packhorses persisted well into the 19th century, creating a legacy of tracks and narrow, stone-arched, packhorse bridges. What started as an industrial necessity involving human and equine labour on an unimaginable scale has become a rambler's delight through some of England's most dramatic and beautiful countryside. And in spite of periodic grumbles about our present-day roads and railways, 'troublesome travel', as understood by the Duke of Bridgewater and his engineer, James Brindley, is thankfully a thing of the past.

A rambler's delight: the packhorse bridge at Wasdale Head, Cumbria (Geograph/Perry Dark).

3 Thomas Telford

A start in life

Looking back on our boating years I have become fascinated by the prodigious feats of engineering that brought the English canals into being, and by the men who designed and built them. What mixture of family background, training and ambition propelled these rare individuals to venture far beyond the limits of existing knowledge and experience, accepting challenges that others pronounced foolish, or dangerous, or both?

Thomas Telford (1757–1834) is probably the greatest civil engineer Britain has ever produced. Coming from the humblest of backgrounds, he taught himself the practical and intellectual tools of his trade without a whiff of university privilege, and went on to design and build pioneering canals, roads, bridges, docks and harbours. Never married, his time was almost totally devoted to his work and therefore – some would claim – lacked the emotional peaks and troughs that accompany most lives. But to delve deeper is, as so often, to reveal a more complex tapestry.

Samuel Smiles, who has already given us valuable insights into 'troublesome travel' in the 18th and 19th centuries, also wrote admirably about the life and times of Thomas Telford. He has a clear view of the engineer's humble beginnings in Eskdale, an isolated valley in Scottish Border country about 20 miles north of Gretna Green:

> One would scarcely have expected to find the birthplace of the builder of the Menai Bridge and other great national works in so obscure a corner of the kingdom. Possibly it may already have struck the reader with surprise, that not only were all the early engineers self-taught in their profession, but they were brought up mostly in remote country places, far from the active life of great towns and cities. But genius is of no locality, and springs alike from the farmhouse, the peasant's hut, or the herd's shieling. Strange, indeed, it is that the men who have built our bridges, docks, lighthouses, canals, and railways, should nearly all have been country-bred boys.

Apart from recent afforestation, the countryside in which Telford was born and bred has changed little, and remains enticingly far from the madding crowd. Thankfully it is served by modern roads and we may get a good feel for it by travelling from Langholm,

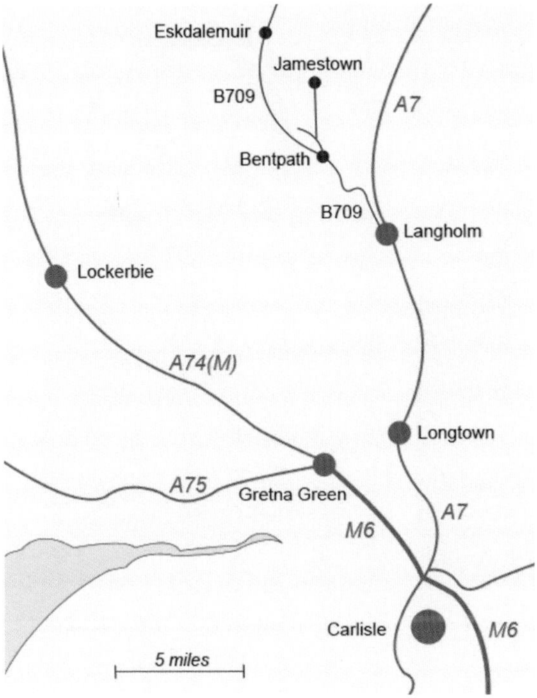

Thomas Telford was born near the hamlet of Jamestown, in deep countryside about 20 miles north of Gretna Green.

the local town, along the B709 through Eskdale. This is gentle country by Scottish standards, more rural idyll than highland drama, with rich green hills sweeping down to valley floors and sparkling streams feeding the River Esk, one of the finest salmon rivers in the country.

After 5 miles we come to Bentpath – larger than a hamlet, smaller than a village – in the expansive rural parish of Westerkirk. But rather than turn right to Westerkirk Bridge and the church, we will follow the main road a little further to discover a surprising building for such a spot: neither chapel nor village school, but a *library*; and one that has been described as the best equipped, given its location and the population it serves, of any in the country. It is also the oldest library still in use in Scotland, and holds over 8,000 books dating from the 18th century onwards, with records that give a fascinating insight into the reading habits of a small rural community. There must be a good story here somewhere.

That story began in 1791 when the Westerhall Mining Company started mining and smelting antimony at Glendinning, just a stone's throw from the cottage where Thomas Telford had entered the world 34 years previously. The company built an access road, bridges and a row of miners' cottages beside Meggat Water; and, much more surprisingly, a library, which completed the settlement of Jamestown. But the Louisa Mine had a short life, closing in 1799 after yielding 100 tons of semi-pure antimony with a value of £8,400. The library was relocated in the Bentpath village school that Thomas had attended, and relocated again in 1840 to a brand new building – the one we see today.

As we shall discover, Thomas became fascinated by books in his teenage years and carried a love of poetry through to the end of his life. It is doubtful whether he ever saw miners at work in the short-lived Louisa Mine because by the 1790s he was engaged far away as a professional engineer, but he must have heard about the library and its subsequent removal to cramped conditions in the village school. When he died in 1834 he bequeathed £1000 'to the minister of Westerkirk in trust for the parish library', a princely sum which grew to £2,700 by the time his estate was settled.

Westerkirk library and the Telford memorial seat near Bentpath (Geograph/Walter Baxter).

Today we see a fine Victorian building and a more recent stone memorial at the roadside. One of its plaques announces:

> This seat was erected in 1928 to perpetuate the memory of Thomas Telford, son of the unblameable shepherd, and to record his fame as an engineer and his untiring benevolence. Apprenticed to a stonemason in Langholm, his creative genius gave to the nation many works of inestimable benefit. He was the first President of the Institution of Civil Engineers.

Another plaque carries lines by the man himself:

> There 'mongst those rocks I'll form a rural seat
> and plant some ivy with its moss compleat
> I'll benches form of fragments from the stone
> which nicely pois'd was by our hands oerthrown.

By the early 1990s the library was in urgent need of TLC. An ambitious five-year project restored the building and its stock of books to their former glory, giving visitors a glimpse of records going back over 200 years. At the time of writing admission is free but opening times are limited. Borrowing of books is restricted to residents of Westerkirk and the adjoining parishes.

We now retrace our steps – in space, a few hundred yards back along the B709 before turning left into the lane that leads over Westerkirk Bridge to Bentpath; and in time, to discover Thomas's birthplace in a miniature valley described by Samuel Smiles as 'one of the most solitary nooks of the narrow valley of the Esk'.

Westerkirk Bridge, built some 20 years before Tom's birth, overlooks the bustling river. He must have wandered over it hundreds of times as a child and teenager, little dreaming that he would become one of the world's great bridge builders. Nearby is Westerkirk parish church (not the building he knew) and the churchyard with a simple headstone that marks the grave of a poor shepherd, embellished with the earliest known example of letter carving by a young man who would soon become a highly skilled stonemason:

IN MEMORY OF
JOHN TELFORD,
WHO AFTER LIVING 33 YEARS
AN UNBLAMEABLE SHEPHERD,
DIED AT GLENDINNING,
NOVEMBER, 1757.

The shepherd was, of course, his father.

Westerkirk Bridge over the River Esk, completed in 1736 (Geograph/Walter Baxter).

The Telford cairn at Glendinning, a stone's throw from Jamestown (Geograph/Walter Baxter).

Almost subconsciously I have started to call the baby and the young lad 'Tom'. Previously, like the majority of his biographers and commentators, I preferred Thomas Telford, or just plain Telford. Yet we are about to discover a birthplace so humble, and a family situation so fraught, that a description of his childhood seems best uncluttered by adult formality. Time enough later for Telford, giant of the civil engineering profession.

The lane bears left at the far end of Westerkirk Bridge to follow the north bank of the Esk. A mile or so later a right turn leads up alongside Meggat Water to Jamestown; and finally we reach Glendinning and a simple stone cairn, erected in 2007 on a knoll where a mud-walled cottage 'little better than a shieling' once stood – the spot where Tom was born among the hills.

The tablet marks the 250th anniversary of Thomas Telford's birth (Geograph/Oliver Dixon).

The setting has changed little in the intervening years:

> From the knoll may be seen miles on miles of hills up and down the valley, winding in and out, sometimes branching off into smaller glens, each with its gurgling rivulet of peaty-brown water flowing down from the mosses above … at Glendinning you seem to have got almost to the world's end. There the road ceases, and above it stretch trackless moors, the solitude of which is broken only by the whimpling sound of the burns on their way to the valley below, the hum of bees gathering honey among the heather, the whirr of a blackcock on the wing, the plaintive cry of ewes at lambing-time, or the sharp bark of the shepherd's dog gathering the flock together for the fauld.

But the rural idyll was accompanied by tragedy. The mother had lost a previous baby Tom, and the happy arrival of a successor was soon overshadowed by dreadful news:

> In this cottage on the knoll Thomas Telford was born on the 9th of August, 1757, and before the year was out he was already an orphan. The shepherd, his father, died in the month of November, and was buried in Westerkirk churchyard, leaving behind him his widow and her only child altogether unprovided for.

Tom's mother, from now on known by her maiden name, Janet Jackson, in line with local custom, was in deep trouble: alone and penniless, living in a tied cottage that must be surrendered now her 'unblameable' husband had passed away, along with an infant of four months to be nursed, fed and clothed. She would face a long hard struggle with enormous fortitude, overcoming problems that would have destroyed lesser spirits. It seems that baby Tom, destined to become an engineer of legendary ambition and determination, inherited a generous assortment of genes from his mother.

We tend to associate dire living conditions in 18th- and 19th-century Britain with the urban poor rather than their country cousins, perhaps above all thanks to the novels of Charles Dickens. The much-loved British countryside in which so many of us enjoy week-end recreation seems an unlikely setting for destitution. How on earth could Glendinning's natural beauty hide poverty to rival an urban slum?

In fact Janet's bereavement, and Tom's entry into the world, took place in a greatly diminished Scotland. It was only 57 years since the Act of Union had joined the parliaments of England and Scotland to form Great Britain, and in the meantime the Highlanders had launched two Jacobite rebellions that threatened what, to most of the English, had seemed an established Protestant order. The second rebellion, led by Bonnie Prince Charlie in 1745, raised an army large enough to take Edinburgh and march as far south as Derby – a profoundly unnerving crisis for the London government. Although subsequent retribution, swift and brutal, was principally aimed at the Highlanders and their clan system, a continuing air of crisis and uncertainty in Scotland reduced the chances that her citizens would feel much improvement in their lives, at least in the short term.

In any case there was a longstanding problem with Scottish infrastructure which, insofar as it existed at all, was in a lamentable state. At the time of the Union in 1707 England had about 5 times the population of Scotland but 36 times the wealth. The few Scottish roads were even worse than those in England, the land more sterile, the people poorer. Small towns and villages were often little more than collections of thatched mud cottages. Communications, always difficult, were often impossible in winter. The remarkable improvements in roads, bridges, buildings and agriculture that took place in the following 100 years would become the main drivers of economic and social progress, finally raising the country from its slough of despond; but at the time of Janet Jackson's bereavement in 1757, that was all for the future – a future to which her infant son would make a massive contribution.

In the meantime most Lowland farms continued as unenclosed 'out-field', typically rough open moorland, and a small enclosed patch of 'in-field' intended for oats or barley but commonly dominated by weeds. The value of manure was so little understood that it was often thrown into a nearby stream to float away, or burned in summer. Most farming families shared living accommodation with their livestock.

According to Samuel Smiles there was still a veritable slave class in Scotland

> who were bought and sold with the estates to which they belonged, as forming part of the stock. When they ran away, they were advertised for as negroes were in the American States … Money was then so scarce that Adam Smith says it was not uncommon for workmen, in certain parts of Scotland, to carry nails instead of pence to the baker's or the alehouse. A middle class could scarcely as yet be said to exist, or any condition between the starving cottiers and the impoverished proprietors, whose available means were principally expended in hard drinking … and were, for the most part, too proud and too ignorant to interest themselves in the improvement of their estates.

Rural Dumfries and Galloway were in a derelict state 'without made roads of any kind'. But fortunately for Janet and her son, the lawlessness that had existed for centuries in the border country between Scotland and England, where 'reivers' rustled livestock, stole household goods, took prisoners for ransom, and even murdered, had subsided since the Act of Union. Extended families, referred to as clans, which had acted as miniature armies and devoted their energies to plunder, were now in more cooperative mood. The Esk valley was among the beneficiaries, and although poverty was still rife, there was, as Smiles observes, less fear and misery:

> The farmers of the dale were very primitive in their manners and habits, and being a warm-hearted, though by no means a demonstrative race, they were kind to the widow and her fatherless boy. They took him by

A raid by reivers on Gilnockie Tower, Dumfries and Galloway (Wikipedia/Cattermole).

> turns to live with them at their houses, and gave his mother occasional employment. In summer she milked the ewes and made hay, and in harvest she went a-shearing; contriving not only to live, but to be cheerful.

So in spite of a total absence of state-funded social security or medical care, Janet survived. In such a small community, little or no stigma attached to accepting a neighbour's help, considered a friendly act that elevated the donor as well as the recipient. She was soon moved about a mile down the valley to a place called The Crooks, and found herself re-housed in another primitive mud-walled thatched cottage, this time with a central doorway and two rooms, one for her and Tom, the other for neighbours called Elliot. From there she brought up an only child – who, when an old man greatly honoured in his own lifetime, declared: 'I still recollect with pride and pleasure my native parish of Westerkirk, on the banks of the Esk, where I was born.'

The child grew into a healthy boy, so full of fun that he became known in the valley as 'Laughing Tam'. When he was old enough to herd sheep he stayed with a relative, a shepherd like his father, and spent much of his time on the hills. In winter he often lodged with neighbouring farmers, herding cows or running errands and receiving in return food, a pair of stockings and wooden clogs.

Time spent alone among the sweeping hills of Eskdale had a profound effect on young Tom, encouraging a deep love of the natural world and his native place. It is surely no coincidence that as an eminent engineer he would execute many of his greatest works in wild areas of the Scottish Highlands and North Wales – works that show a great sensitivity to landscape and the grandeur of their setting.

Education was also on the agenda. Westerkirk was fortunate to have a parish school, representative of an admirable institution that had developed at an early date in Scotland. The fees, although modest, were beyond Janet's means, so she reluctantly accepted help from her brother – help that proved of enormous benefit:

> By imparting the rudiments of knowledge to all, the parish schools of the country placed the children of the peasantry on a more equal footing with the children of the rich; and to that extent addressed the inequalities of fortune. To start a poor boy on the road of life without instruction, is like starting one on a race with his eyes bandaged or his leg tied up … It was not much that he learnt … but there was another manifest advantage to the poor boy in mixing freely at the parish school with the sons of the neighbouring farmers and proprietors. Such intercourse has an influence on a youth's temper, manners, and tastes, which is quite as important in the education of character as the lessons of the master himself; and Telford often, in after life, referred with pleasure to the benefits which he had derived from his early school friendships.

Prominent among his friends was Andrew Little, who went on to study medicine in Edinburgh and later enrolled as ship's doctor on an African slaving vessel. The voyage was a disaster, and not only for the slaves, because lightning savaged the ship during a storm and left the young doctor permanently blinded. His career in medicine cruelly truncated, he returned to Langholm and became village schoolmaster. When Thomas subsequently left Eskdale for London, he started a long correspondence with his blind friend and was always delighted to reminisce about their schooldays, joking about 'the glutton boy' John Elliot, and enquiring about Jennie Smith who seems to have taken his fancy – 'tell her she is a canterrin sort of lassie'.

But what would Tom do on leaving school? The 15-year-old son of a single mother living in a thatched mud cottage in a remote valley had few strings to pull, and the prospects for gainful employment must have looked bleak. Was he to become a shepherd like his father, or a farm labourer, or perhaps a trade apprentice? Local trade was largely confined to building stone walls using skills that 'an ordinarily neat-handed labourer could manage' – not an inspiring prospect – and jobs in the rest of Eskdale were few and far between.

When Samuel Smiles visited the area many years later he asked a local what would happen to the many children he saw in local villages, and received a blunt answer:

They swarm off! If they remained at home, we should all be sunk in poverty, scrambling with each other amongst these hills for a bare living. But our peasantry have a spirit above that: they will not be content to sink; they look up; and our parish schools give them a power of making their way in the world, each man for himself. So they swarm off – some to America, some to Australia, some to India, and some, like Telford, work their way across the border and up to London.

The quotation confirms that this member of the local 'peasantry' had sufficient spirit to 'swarm off'. But it jumps the gun on our story, because another nine years would pass before Thomas Telford took the road to London. His flight from Eskdale was to take place in several distinct hops.

The first was to the village of Lochmaben, about 20 miles from home, where he started an apprenticeship with a stonemason in 1772. But his new master ill-treated him – we have no details – and a few months later Laughing Tam was back with his mother at The Crooks. She must have been worried and disappointed, but luckily her nephew Thomas Jackson, who had recently been appointed land steward to the Johnstones, a wealthy local family, stepped in to help. As a man with some influence, Jackson was able to persuade Andrew Thomson, a well-respected mason in Langholm, to take on Thomas for the remainder of his apprenticeship.

So his second hop was to Langholm, and it benefited the future civil engineer enormously. Today's Langholm is a small neat town of about 2,500 people, well known for its woollen textile industry and as the historic centre of the Armstrong clan that had made its own notable contribution to cross-border raids before the Act of Union. But back in 1772 it was a village consisting chiefly of 'mud hovels' and 'no better than the district that surrounded it'. Fortunately, it was about to undergo major improvements, thanks to the young Duke of Buccleuch, who had just succeeded to his family's local estates. The Buccleuchs, like the Armstrongs, were a prime example of a family that had thrown its energy into border raids for centuries, but were now redirecting it, somewhat tardily, to less destructive activities:

The energy which the old borderers threw into their feuds has not become extinct, but survives under more benignant aspects, exhibiting itself in efforts to enlighten, fertilize, and enrich the country which their wasteful ardour before did so much to disturb and impoverish. The heads of the Buccleugh and Elliot family now sit in the British House of Lords … The border men, who used to make such furious raids and forays, have now come to regard each other, across the imaginary line which divides them, as friends and neighbours; and they meet as competitors for victory only at agricultural meetings, where they strive to win prizes for the biggest turnips or the most effective reaping machines.

It is clear that Thomas, who had entered the world with almost no prospect of advancement, now found himself in an environment that was changing rapidly for the better. In this he was lucky: born 50 years earlier, the seeds of his ambition might well have fallen on stony ground; 50 years later, on fertile soil already claimed by other talented individuals. As it was, the Duke of Buccleuch's programme of works in Langholm – paving rough tracks, bridging fords, building stone houses in place of mud and thatch – produced a demand for masons' work at just the right time for Thomas, and introduced him to a wide variety of skills under his new master, Andrew Thomson.

Langholm had grown up on the east bank of the Esk, but by the time Thomas completed his apprenticeship it was extending westwards. He helped build houses in the New Town, some with ornamental doorways that demonstrated his increasing confidence as a craftsman in stone and the pride he took in his work. But above all it was a new bridge thrown over the Esk by a master mason, Robin Hotson, in cooperation with Andrew Thomson that introduced him to the massive structures that would play such a big part in his professional life. Many of its stones were hewn by Thomas's hand and carried his personal mason's mark – a vertical line with an arrowhead at the top and a lozenge at the bottom, separated by a cross. When the Esk runs low in summer some of the marks may still be detected below the western arch.

Soon after the bridge was completed in 1778 an unusually high flood surged down the Esk valley. The river was 'roaring red frae bank to brae' and onlookers feared the structure would be swept away. Hotson was away from home at the time and his wife, Tibby, knowing he was under contract to maintain the bridge for seven years, panicked. She ran sobbing from one person to another, calling out 'Where's Tammy Telfer, where's Tammy? … they say it's shakin, it'll be doon.' In spite of Tammy's reassurances and evident amusement, she insisted the bridge was shifting and ran up to stiffen her back against the parapet apparently in order to support the bridge with her own body. Needless to say, she need not have bothered; the bridge stands firm to this day.

Thomas worked in Langholm and Eskdale for eight years, latterly as a journeyman mason in his own right. He built farmhouses on the Duke of Buccleuch's estate as well as the Manse at Westerkirk. As work began to slacken off in Langholm he undertook small jobs, including the hewing of gravestones and ornamental doorways. Throughout this period he kept in touch with his mother and often visited her at The Crooks on Saturday evenings, accompanied her to the parish church on Sundays, and displayed a compassion that would endure long after he became famous. But he had now learned all that his native valley could teach him in the art of masonry. In 1780 he decided to leave Eskdale and seek work 50 miles away in Edinburgh.

Thomas's years in Eskdale introduced him to arts other than stonemasonry, and none more significant for his personal development than the riches of English literature. This came about in a surprising way. An elderly lady, Miss Pasley, who lived in one of the few smart houses in Langholm, became aware of his family situation:

> As the town was so small that everybody in it knew everybody else, the ruddy-cheeked, laughing mason's apprentice soon became

Langholm Bridge over the River Esk, worked on by Thomas Telford. Old Langholm is on the right, New Town on the left (Geograph/John Chroston).

> generally known to all the townspeople, and amongst others to Miss Pasley. When she heard that he was the poor orphan boy from up the valley, the son of the hard-working widow woman, Janet Jackson, her heart warmed to the mason's apprentice, and she sent for him to her house. That was a proud day for Tom; and when he called upon her, he was not more pleased with Miss Pasley's kindness than delighted at the sight of her little library of books, which contained more volumes than he had ever seen before.

He had already borrowed, and exhausted, the few books owned by his young friends, so was delighted when Miss Pasley offered him volumes from her own collection. From now on he was almost never without a book, to be read during snatched intervals in his working day, among his beloved hills, or on winter evenings by the uncertain light of a candle or cottage fire.

On one occasion Miss Pasley lent him Milton's *Paradise Lost*. Its ten thousand lines of verse covering the greatest of all themes, the struggle between good and evil, intoxicated him: 'I read and read, and glowred; then read and read again.' In a young man of 20 who was spending most of his time fashioning stone in the company of people with little formal education, his sensitivity to the rhythmic power of John Milton's epic poem seems remarkable. It probably stirred feelings experienced among the wilds of nature, but never so far expressed.

Thomas went on to read contemporary poets, including Robert Southey (1774–1843), who became Poet Laureate and accompanied him on a tour of the Scottish Highlands many years later. He also began to write verse of his own – usually, as he admitted, not very well: 'It is, to me, something like what a fiddle is to others, I apply to it in order to relieve the mind after being much fatigued with close attention to business.' Nevertheless some of his efforts, including lines addressed to his hero, Robert Burns, and verses in praise of the hill countryside where he was born, achieved publication. The most ambitious of his early efforts was 'Eskdale', which prompted Southey to declare that 'many poems which evinced less observation, less feeling, and were in all respects of less promise, have obtained university prizes'.

Discovery of poetry and literature awakened Thomas to a wider world of ideas and helped him feel at ease in educated company. He became such a competent writer that he was often asked by acquaintances to pen letters on their behalf:

> One evening a Langholm man asked Tom to write a letter for him to his son in England; and when the young scribe read over what had been written to the old man's dictation, the latter, at the end of almost every sentence, exclaimed, "Capital!, capital!" and at the close he said, "Well! I declare, Tom! Werricht himsel' couldna ha' written a better!" – Wright being a well-known lawyer or "writer" in Langholm.

Thomas's hop to Edinburgh introduced him to a New Town vastly more impressive than Langholm's. Princes Street was 'rising as if by magic', and skilled masons were in great

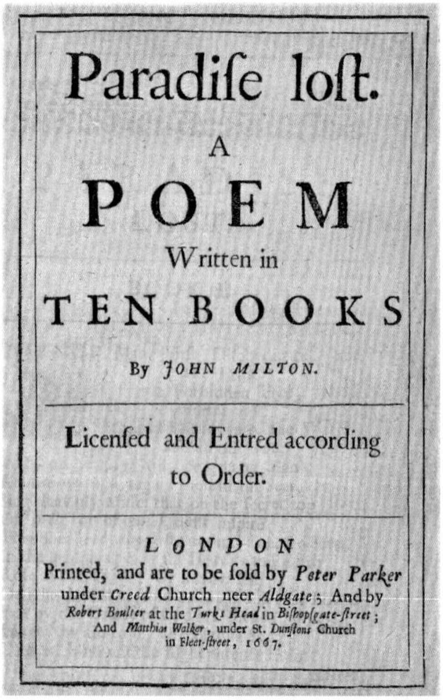
Paradiſe loſt.
A
POEM
Written in
TEN BOOKS
By JOHN MILTON.
Licenſed and Entred according to Order.
LONDON
Printed, and are to be ſold by *Peter Parker* under *Creed* Church neer *Aldgate*; And by *Robert Boulter* at the *Turks Head* in *Biſhopſgate-ſtreet*; And *Matthias Walker*, under St. *Dunſtons* Church in *Fleet-ſtreet*, 1667.

Of man's first disobedience, and the fruit

Of that forbidden tree, whose mortal taste

Brought death into the world, and all our woe ...

John Milton's epic poem Paradise Lost was first published in 1667 (Wikipedia).

Feasts for Thomas Telford's eyes: above, Edinburgh Castle (Geograph/Richard Croft) and Holyrood House (Geograph/John Lord); below: Heriot's Hospital, Edinburgh (Wikipedia/ Oliver Bonjoch) and Melrose Abbey (Geograph/James Barton).

demand. He arrived at a time of transformation not only in city layout and architecture but in science, medicine, industry, agriculture, political economy and philosophy, the fruits of Scotland's gathering Age of Enlightenment. New ideas based on evidence and reasoned argument were challenging traditional authority, especially religious authority, in the optimistic belief that humans could improve society without divine help – a philosophical stance similar, in our time, to that of humanism. Its intellectual climate probably suited Thomas Telford's personality down to the ground.

He remained in Edinburgh for about 18 months:

> during which he had the advantage of taking part in first-rate work and maintaining himself comfortably, while he devoted much of his spare time to drawing, in its application to architecture. He took the opportunity of visiting and carefully studying the fine specimens of ancient work at Holyrood House and Chapel, the Castle, Heriot's Hospital, and the numerous curious illustrations of middle age domestic architecture with which the Old Town abounds.

Thomas was clearly aiming high – indeed, so high that even a buoyant Edinburgh could not contain him: 'Having acquired the rudiments of my profession, I considered that my native country afforded few opportunities of exercising it to any extent, and therefore judged it advisable (like many of my countrymen) to proceed southward, where industry might find more employment and be better remunerated.' After a final architectural visit to Melrose Abbey, a former Cistercian monastery in Border country about 40 miles southeast of Edinburgh, he returned to Eskdale to say goodbye to his old contacts, including:

> the neighbouring farmers, who had befriended him and his mother when struggling with poverty – his schoolfellows, many of whom were preparing to migrate, like himself, from their native valley – and the many friends and acquaintances he had made while working as a mason in Langholm. Everybody knew that Tom was going south, and all wished him God speed.

Yet it seems surprising that Edinburgh's burgeoning New Town failed to satisfy his ambition, at least for a few more years, because it was becoming a showcase for the neoclassical architecture in vogue in England. The city fathers, realising that the hopelessly overcrowded Old Town no longer suited its growing professional and merchant classes, had sanctioned building of the first phase of New Town in 1767, 13 years before Thomas's arrival. It was to continue until 1820. A fine example of its new architecture is Dundas House in St Andrew Square, which must surely have delighted Thomas's eye. It is hard to believe that the talented stonemason exhausted Edinburgh's possibilities so rapidly.

Dundas House in Edinburgh's New Town, completed in 1774, became headquarters of the Royal Bank of Scotland in 1821 (Wikipedia/Thunderwing).

Actually we do not really know the details of Thomas's stay in Edinburgh. He himself was silent on the subject. Samuel Smiles offers no clues. Perhaps the 24-year-old encountered personal difficulties with bosses or colleagues – or had met, and fallen for, a young woman who jilted him. In any case his mind was made up, and late in 1781 he embarked on his next adventure: like Dick Whittington, but without the cat, he set off for London.

To London, Portsmouth and Shropshire

> A common working man, whose sole property consisted in his mallet and chisels, his leathern apron and his industry, might not seem to amount to much in "the great world of London". But, as Telford afterwards used to say, very much depends on whether the man has got a head with brains in it of the right sort upon his shoulders. In London, the weak man is simply a unit added to the vast floating crowd, and may be driven hither and thither, if he do not sink altogether; while the strong man will strike out, keep his head above water, and make a course for himself, as Telford did.
>
> There is indeed a wonderful impartiality about London. When work of importance is required, nobody cares to ask where the man who can do it best comes from, or what he has been, but what he is, and what he can do. Nor did it ever stand in Telford's way that his father had been a poor shepherd in Eskdale, and that he himself had begun his London career by working for weekly wages with a mallet and chisel.

So how did Thomas Telford and his hand tools get to London? His early years among the hills of Eskdale had made him a great walker, and he thought nothing of 20 miles a day – but it was 350 miles to the capital. He had learned 'rough riding' as a youth, but did not own a horse. Travel by stage coach was certainly improving in the late 1700s, but the journey from Edinburgh to London meant staying in coaching inns and that, combined with the stagecoach fare, was beyond the means of a journeyman mason. None of these fitted the bill.

Thomas Telford as a young man, a portrait by Scottish painter Henry Raeburn (ICE).

By chance his cousin Thomas Jackson came to the rescue once more. Jackson's employer, Sir James Johnstone of Wester Hall, had promised a horse to a family member in London but was unable to find a suitable rider. Not only did Jackson secure the invitation, but he lent Thomas Telford his buckskin breeches to cushion him against the hardships of the road. The young mason left for London well mounted, and delivered Sir James's horse without incident. Jackson often told the story of his cousin's first ride to London with considerable relish, taking care to wind up with 'but Tam forgot to send me back my breeks!'

Thomas's luck continued. Before leaving Langholm he had been asked by his elderly friend, Miss Pasley, to deliver a letter to her brother John, a wealthy London merchant. As a result the young stonemason obtained an introduction to Sir William Chambers, architect of one of London's most famous neoclassical buildings, Somerset House, then under construction. It was 'the finest architectural work in progress in the metropolis' and Thomas wanted to be part of it. From the start he was determined to demonstrate his skills in fine carving, rather than the mere fashioning of solid blocks, and was soon being employed as a first-class mason. He had landed on his feet and, to judge by letters sent to friends in Eskdale, was happy, cheerful and keen to receive news of those he had left behind. The softer side of his personality surfaced in a letter written after more than a year's absence, envying the planned visit to Eskdale by a young surgeon acquaintance – 'for the meeting of long absent friends is a pleasure to be equalled by few other enjoyments here below'.

He worked with many fellow masons during his time at Somerset House and his 'Laughing Tam' personality stood him in good stead. Yet quiet observation convinced him that most of his companions lacked spirit and, above all, forethought:

> He found very clever workmen about him with no idea whatever beyond their week's wages. For these they would make every effort: they would work hard, exert themselves to keep their earnings up to the highest point, and very readily "strike" to secure an advance; but as for making a provision for the next week, or the next year, he thought them exceedingly thoughtless. On the Monday mornings they began "clean", and on Saturdays their week's earnings were spent.

Unsurprisingly, his own approach was entirely different:

Somerset House, London, soon after completion, showing its frontage onto the River Thames, and its façade onto the Strand (Wikipedia).

> Telford, on the other hand, looked upon the week as only one of the storeys of a building; and upon the succession of weeks, running on through years, he thought that the complete life structure should be built up.

Yet again we get the impression of a young man who despite the humblest of beginnings harboured an extraordinary ambition; every work experience must be put to good use, every shilling earned was means to an end, not an end in itself.

The most talented of his fellow masons, a man called Hatton, was 'honesty and good nature itself' and the only one to become a close friend. He had been working on Somerset House for six years and was 'esteemed the finest workman in London, and consequently in England'. With a thorough understanding of architectural drawings, he worked equally well in stone and marble, cutting Corinthian capitals and other elaborate ornaments with confidence. But he, too, was content to work as a journeyman, earning just a few shillings a week above average wages.

Thomas Telford suggested to his friend that they set up in business together, noting that there was 'nothing done in stone or marble that we cannot do in the completest manner'. Robert Adam, one of the most famous neoclassical architects of the age and an enthusiast of the Grand Tour, learned of their plans and promised support. But the enterprise was not to be: their great difficulty was lack of business capital. Thomas eventually had to accept it as an insuperable hurdle and with great regret abandoned the scheme; but it was an indication of a growing determination to become his own master.

Perhaps that is what led to his next major hop. In 1784, at the age of 27, he obtained a contract to build a fine residence for the Commissioner of Portsmouth Naval Dockyard, designed by another well-known neoclassical architect, Samuel Wyatt. The contacts that Telford had made in London, and especially his work at Somerset House, were bearing fruit. The contract included several other buildings and a new chapel (now St Ann's Church), giving him the chance to roam among the dockyard's many facilities, including the massive Great Stone Dock of 1689, which had recently been upgraded.

It is also likely that he saw HMS *Victory*, either berthed in the dockyard or sailing close offshore in the Solent or Spithead. Commissioned in 1778, she had already seen three naval engagements, the most recent off Gibraltar in 1782, and was destined to become Nelson's flagship at Trafalgar in 1805. Telford must have been aware of her growing reputation as a formidable first-rate ship of the line, and inspired by yet another example of Britain's growing industrial and naval pre-eminence. Today, she is a world-famous resident

The Commissioner's House in Portsmouth Naval Dockyard, built by Telford (Wikipedia).

HMS Victory *in permanent dry dock in Portsmouth (Wikipedia/Ballista).*

of Portsmouth Dockyard, a major tourist attraction, and the world's oldest naval vessel still in commission.

Thomas Telford's buildings progressed so satisfactorily that the Commissioners started to value his advice above that of the dockyard superintendent – 'a dangerous point, being difficult to keep their good graces as well as his. However I will contrive to manage it.' His daily schedule was tightly controlled: rising at 7 in the morning in winter, 5 in summer, he worked on business matters until breakfast at 9; spent the rest of the morning among his buildings, checking progress and giving orders; lunched at 2, followed by writing, drawing and reading until 9.30; then supper and bed.

> This is my ordinary round, unless when I dine or spend an evening with a friend; but I do not make many friends, being very particular, nay, nice to a degree. My business requires a great deal of writing and drawing ... Then, as knowledge is my most ardent pursuit, a thousand things occur which call for investigation which would pass unnoticed by those who are content to trudge only in the beaten path. I am not contented unless I can give a reason for every particular method of practice which is pursued. Hence I am now very deep in chemistry ... I have borrowed a MS copy of Dr Black's lectures. I have bought his "Experiments on Magnesia and Quicklime", and also Fourcroy's Lectures, translated from the French by one Mr Elliot, of Edinburgh. And I am determined to study the subject with unwearied attention

until I attain some accurate knowledge of chemistry, which is of no less use in the practice of the arts than it is in that of medicine.

Such words are crucial for our understanding of Thomas Telford, whose formal education had stopped when he left the parish school in Westerkirk. He had received no help from high school or university; now, at the age of 27, he was immersing himself in the mysteries of chemistry, a prelude to a lifelong course of self-education. The following year he wrote to his blind friend, Andrew Little, in Langholm that he was 'chemistry mad', and that the pocket book which he always carried was becoming crammed with 'facts relating to mechanics, hydrostatics, pneumatics, and all manner of stuff'.

But in case we should think him a narrow workaholic, he assures us that he is taking a great delight in Freemasonry, and is about to have a room decorated and furnished to his own design at the George Inn, where he has his hair powdered every day and puts on a clean shirt three times a week. Not that attention to personal appearance denotes conceit – he would 'rather have it said of him that he possessed one grain of good nature or good sense than shine the finest puppet in Christendom'. And no matter how much he was burdened with work, he continued to write letters home, especially to his mother in The Crooks, using block capitals so that she could decipher them by her cottage fireside. As Samuel Smiles remarks:

As a man's real disposition usually displays itself most strikingly in small matters – like light, which gleams the most brightly when seen through narrow chinks – it will probably be admitted that this trait, trifling though it may appear, was truly characteristic of the simple and affectionate nature of the hero of our story.

Thomas completed his buildings in Portsmouth Dockyard in late 1786, and his contract came to an end. There was not much else to detain him in Portsmouth – but where should he go next? Back to London, where neoclassical architecture was in full flow? Or perhaps to Edinburgh, where his newly acquired knowledge and experience of prestigious buildings would surely open new doors? But as it happened his next opportunity was offered by another contact from the past: William Pulteney, who had moved away from Eskdale many years before and was by now the highly influential Member of Parliament for Shrewsbury, county town of Shropshire.

Pulteney had started life as William Johnstone, the second son of Sir James Johnstone of Wester Hall in Eskdale (the man who had taken on Thomas's cousin as land steward and entrusted one of his horses to Thomas for his ride to London). William studied law and became an eminent advocate in Edinburgh, where he became a friend of Enlightenment figures including architect Robert Adam, philosopher David Hume and economist Adam Smith. In 1760 he married the heiress Frances Pulteney, a cousin of the 1st Earl of Bath, with two major consequences: he changed his name from Johnstone to Pulteney; and he inherited fabulous wealth, including large estates in Shropshire. He went on to invest in lands in North America and many developments in Britain including the famous Pulteney Bridge in neoclassical Bath, one of only four bridges in the world to have shops on both

William Pulteney (1729–1805) by Thomas Gainsborough; and Pulteney Bridge across the River Avon in Bath, designed by Robert Adam and completed in 1774 (Wikipedia).

sides. And as if that wasn't enough to keep him occupied and amused, he entered politics and became an MP in seven successive Parliaments, first representing Cromarty in Scotland, and then Shrewsbury, which he served for 30 years.

Recalling that Thomas Telford had previously given advice on repairs to the Johnstone mansion in Eskdale, and having decided to restore Shrewsbury Castle as a residence for himself, Pulteney contacted Thomas in 1787 and asked him to superintend alterations designed by Robert Adam. Living quarters were found in the castle for Thomas, and

Shrewsbury Castle (Geograph/David Smith); and Laura's Tower, built by Telford (Geograph/Jeremy Bolwell).

Two of Shrewsbury's famous 17th-century timber buildings, Castle Gates House and Council House Gateway (Geograph/Humphrey Bolton).

while there he found time to build a delightful tower in the grounds as a summerhouse for Pulteney's daughter, Laura. Soon afterwards, no doubt thanks to Pulteney's influence, he was elevated to Surveyor of Public Works for the county of Shropshire. It marked the start of a fruitful collaboration and friendship – so much so that he became known in Shrewsbury as 'Young Pulteney'.

As county surveyor, Thomas had to familiarise himself with a wide variety of buildings and structures, some new, and others needing inspection, repair or modification. Shrewsbury is famous for its ancient timber buildings, which must have delighted his eye and challenged any assumption that marble and stone were the only fine building materials. Shropshire expanded his horizons yet again – and increasingly as a designer in his own right rather than overseer of other men's plans. By the summer of 1788 he had more than ten jobs in hand, including bridges, roads, drainage works, an infirmary and a new county gaol.

The gaol project is especially interesting because it introduced Telford to one of the most remarkable men of the time, and awakened him to social issues that had so far escaped his concern. In a letter to his Langholm friend, Andrew Little, he reported:

> About ten days ago I had a visit from the celebrated John Howard, Esq. I say I, for he was on his tour of gaols and infirmaries; and those of Shrewsbury being both under my direction, this was, of course, the

The entrance to former Shrewsbury Gaol is little altered from Telford's original design. The niche above the main doorway contains a bust of John Howard (Geograph/Mark Evison).

> cause of my being thus distinguished … I accompanied him through the infirmary and the gaol. I showed him the plans of the proposed new buildings, and had much conversation with him on both subjects … You may easily conceive how I enjoyed the conversation of this truly good man, and how much I would strive to possess his good opinion. I regard him as the guardian angel of the miserable. He travels into all parts of Europe with the sole object of doing good, merely for its own sake, and not for the sake of men's praise.

John Howard had been born in east London in 1726, the son of a successful upholsterer. On his father's death he inherited considerable wealth and settled on an estate in Bedfordshire. In 1773 he was appointed High Sheriff of Bedfordshire and became responsible for its county jail. He was shocked to discover that jailers in English prisons were not salaried but lived off fees paid by prisoners for food, bedding and other necessities. Penniless prisoners often lived in terrible conditions, and demands for payment before release meant that some stayed in jail long after serving their sentences.

Howard's concerns reached Parliament and launched two 1774 reform acts. But fearing that they were being ignored, he embarked on tours of prisons in England, Scotland and Ireland, followed by France, Holland, Flanders, several German states and Switzerland. Over the next few years he added Denmark, Sweden, Spain, Portugal and Russia to the list. At a time when travel was generally troublesome and frequently dangerous, he covered nearly 50,000 miles at his own expense, making seven major journeys between 1775 and 1790. Tragically for a man of such profound humanitarian instincts, he contracted typhus while visiting military hospitals in Ukraine, and died there in January 1790. The Howard

League for Penal Reform was founded 76 years later in his honour – and continues its work today as the oldest penal reform charity in the UK.

So when John Howard visited Shrewsbury and met Thomas Telford, he had already devoted 15 years of his life to improving the treatment of prisoners at home and abroad. Not that Thomas's reverence for the hugely experienced campaigner spared him criticism: plans for the infirmary and gaol had to be substantially modified, the latter because Howard considered 'the interior courts too small, and not sufficiently ventilated'. But Thomas was happy to follow his guidance, not least because he realised that his new hero was physically and emotionally exhausted:

> he assures me that he hates travelling, and was born to be a domestic man. He never sees his country-house but he says within himself, "Oh! might I but rest here, and never more travel three miles from home; then should I be happy indeed!" But he has become so committed, and so pledged himself to his own conscience to carry out his great work, that he says he is doubtful whether he will ever be able to attain the desire of his heart – life at home. He never dines out, and scarcely takes time to dine at all: he says he is growing old, and has no time to lose. His manner is simplicity itself. Indeed I have never yet met so noble a being. He is going abroad again shortly on one of his long tours of mercy.

As Thomas may have guessed, and John Howard probably knew, that tour of mercy would be his last. He died at Kherson, on the shores of the Black Sea, less than two years after meeting his young admirer in Shrewsbury.

Another formative experience for Thomas Telford in 1788 was the collapse of St Chad's Church in Shrewsbury. Four centuries old, it had been in a lamentable state for some time, with a roof that directed rain onto the congregation. Thomas was called in, and after a rapid inspection of the structure announced to the churchwardens: 'Gentlemen, we'll consult together on the outside, if you please.' He was in no doubt that the building was in a dangerous condition and needed far more extensive repairs than a patched-up roof.

He now had a dispute on his hands. His damning report was rejected by the churchmen on the basis that 'professional men always wanted to carve out employment for themselves, and the whole of the necessary repairs could be done at a comparatively small expense'. A mason in the town was called in to cut away part of a stone pillar, and underpin it. Three days later, as the church bell struck four, its vibrations brought down the tower and demolished the nave, forming, as Telford put it, 'a very remarkable ruin, which astonished and surprised the vestry, and roused them from their infatuation, though they have not yet recovered from the shock'. No doubt the affair did his reputation a power of good in the town; but a brand new St Chad's Church was quickly commissioned and built without any involvement by the county surveyor.

In the same year some extensive Roman remains were unearthed at Wroxeter, a village about 5 miles from Shrewsbury. Viroconium, also known as Uriconium, is believed to have

Among the remains of Viroconium (Geograph/Jeff Buck).

been a major Romano-British city, with a population that grew to 15,000, making it the fourth city of Britain at the time. Samuel Smiles notes that:

> The situation of the place is extremely beautiful, the River Severn flowing along its western margin, and forming a barrier against what were once the hostile districts of West Britain. For many centuries the dead city had slept under the irregular mounds of earth which covered it, like those of Mossul and Nineveh. Farmers raised heavy crops of turnips and grain from the surface and they scarcely ever ploughed or harrowed the ground without turning up Roman coins or pieces of pottery … In fact, the place came to be regarded in the light of a quarry, rich in ready-worked materials for building purposes.

Curiosity was rekindled in 1788 when 'ready worked materials' were needed for a new blacksmith's workshop in the village. As building labourers dug down they disturbed some particularly interesting remains. Antiquarians arrived to inspect and advise, pronouncing the find to be part of a Roman bath house in a remarkable state of preservation; and William Pulteney, who counted the Lordship of the Manor among his many possessions, asked Thomas Telford to take charge of the necessary excavations. An extensive hypocaust (underfloor heating system) was uncovered, together with, as Smiles put it, 'baths, sudatorium, dressing room, and a number of tile pillars – all forming parts of a Roman floor – sufficiently perfect to show the manner in which the building had been constructed and used'. Today the site is owned by English Heritage, and the remains of the bath house are open to the public, together with a visitor centre, shop and museum.

As Thomas moved away from his original passion for stonemasonry towards wider responsibilities, he became increasingly involved in problems of human relationships that beset all management. We have already seen that he decided to tread carefully in Portsmouth to avoid conflict with the dockyard superintendent; to submit, albeit willingly, to design changes requested by John Howard; and to take on the stubborn churchmen of St Chad's. And now, for the first time, he found himself involved with a criminal fraternity outside, rather than inside, a county gaol. As Samuel Smiles notes:

> Among Telford's less agreeable duties about the same time was that of keeping the felons at work. He had to devise the ways and means of employing them without the risk of their escaping, which gave him much trouble and anxiety. "Really", he said, "my felons are a very troublesome family. I have had a great deal of plague from them, and I have not yet got things quite in the train that I could wish. I have had a dress made for them of white and brown cloth, in such a way that they are pye-bald. They have each a light chain about one leg. Their allowance in food is a penny loaf and a halfpenny worth of cheese for breakfast; a penny loaf, a quart of soup, and half a pound of meat for dinner; and a penny loaf and a halfpenny worth of cheese for supper; so that they have meat and clothes at all events. I employ them in removing earth, serving masons or bricklayers, or in any common labouring work on which they can be employed; during which time, of course, I have them strictly watched."

As an antidote we are offered some lighter fare – Thomas's tastes in theatre and music:

> Much more pleasant was his first sight of Mrs Jordan at the Shrewsbury theatre, where he seems to have been worked up to a pitch of rapturous enjoyment. She played for six nights there at the race time, during which there were various other entertainments … With the concert he was completely disappointed, and he then became convinced that he had no ear for music … "the melody of sound is thrown away upon me. One look, one word of Mrs Jordan, has more effect upon me than all the fiddlers in England … I felt no emotion whatever, excepting only a strong inclination to go to sleep … I suppose my ignorance of the subject, and the want of musical experience in my youth, may be the cause of it."

Famous fascinations of Thomas Telford: Dorothea Jordan and Tom Paine (Wikipedia).

So now we know: John Milton and *Paradise Lost*, yes; Shrewsbury theatre and Mrs Jordan, certainly; but music of any kind, emphatically no – the rustle of wind in the trees of Eskdale was more to his taste.

Mrs Jordan (1761–1816) was born in Ireland to an actress mother and a stagehand father who walked out when she was 13. Her impoverished mother saw thespian potential in young Dorothea and put her on the stage. By the age of 21 she was appearing in towns in Yorkshire and, 'having made her escape across the Irish Sea', started calling herself Mrs Jordan. She was attracted to London's theatreland in 1785, and became known at Drury Lane for a special talent in comedy and 'the most beautiful legs ever seen on the stage'. By the time Thomas Telford saw her in Shrewsbury she was 27 years old and one of the most famous actresses of the day.

What he and other besotted admirers may not have known was that Mrs Jordan had already had five affairs with assorted lovers and produced several illegitimate children. They could not have known that the pretty, witty and intelligent actress would soon move upmarket to become mistress of William, Duke of Clarence, later King William IV. The two lived together more or less openly for 20 years and had at least 10 children; but when separation occurred in 1811 she was given an annual allowance and custody of their daughters while he kept custody of the sons. Dorothea was forbidden to return to the stage, but transgressed in 1814 to help pay off the debt of a son-in-law, angering the Duke so much that he took back the daughters and stopped her allowance. She fled to Paris to avoid creditors and died in poverty a year later at the age of 55.

Another famous figure who fascinated Telford at about this time was Tom Paine (1737–1809), English-American political activist, philosopher and revolutionary. Born in

Norfolk, he emigrated to the British American colonies in 1774 and wrote the powerful, widely read pamphlet *Common Sense* to promote the cause of American independence. When the French Revolution broke out in 1789 he transferred attention to France and produced his most famous polemic, *Rights of Man*, in support of the revolutionaries and against their appalled critics in the rest of Europe:

> In the spring of 1791 the first part of Paine's "Rights of Man" appeared, and Telford, like many others, read it, and was at once carried away by it. Only a short time before, he had admitted with truth that he knew nothing of politics; but no sooner had he read Paine than he felt completely enlightened. He now suddenly discovered how much reason he and everybody else in England had for being miserable … Mr Paine had filled his imagination with the idea that England was nothing but a nation of bondmen and aristocrats. To his natural mind, the kingdom had appeared to be one in which a man had pretty fair play, could think and speak, and do the thing he would – tolerably happy, tolerably prosperous, and enjoying many blessings … No one had hindered him; his personal liberty had never been interfered with; and he had freely employed his earnings as he thought proper. But now the whole thing appeared a delusion.

Unsurprisingly, it produced emotional turmoil. He had always accepted the social status quo – indeed had never given it much thought – in spite of being born in poverty and having to haul himself up by his own bootstraps. But he was now the respected County Surveyor of Shropshire, thanks, at least in part, to the patronage of William Pulteney, that extremely wealthy aristocrat. To turn that on its head must have taken quite a bit of mental manoeuvring.

Yet he seems to have managed it, at least temporarily. All of a sudden he had not the slightest difficulty offering revolutionary opinions about county magistrates, MPs and the toffs who were 'carrying the country headlong to ruin'. With Paine's book as a guide, he felt competent to pronounce instant judgement on constitutional arrangements that had lasted for centuries. As he wrote to his Langholm friend, Andrew Little:

> I am convinced that the situation of Great Britain is such that nothing short of some signal revolution can prevent her from sinking into bankruptcy, slavery, and insignificancy.

In the same letter he referred to 'the disgraceful riots' in Birmingham, and the destruction of the home belonging to scientist and clergyman Joseph Priestley. As this was the work of the mob, he could hardly blame the aristocracy; so he sidestepped neatly and blamed the clergy, who had even less to do with it, and wound up with 'may the Lord mend their hearts and lessen their incomes!'

His dramatic change of heart was kept from the townspeople of Shrewsbury, the clergy of St Chad's, and 'the rosy-cheeked old country gentlemen' on whose goodwill, as county surveyor, he depended. But he had a near scrape with the man who had done most to help him since leaving Portsmouth – his patron, and increasingly his friend, William Pulteney.

As an MP, Pulteney enjoyed 'franking privilege' with the Post Office, allowing him to send personal mail free of charge. The system was widely abused, and it was generally expected that anyone with a Parliamentary connection would get his friends' mail franked. But somehow Telford managed to use Pulteney's frank undetected, and posted a copy of Paine's *Rights of Man* to Andrew Little in sleepy Langholm, where it created a sensation. Some of the inhabitants celebrated by drinking revolutionary toasts – and spent the next six weeks behind bars.

Pulteney was indignant when he discovered Telford's deception, but decided to forgive and forget. Telford, willing to forgive at least one of England's aristocrats, and increasingly aware that the liberty won by Parisians was turning to terror and bloodshed, became 'wonderfully reconciled' to the British Constitution and the freedoms it gave him. From now on he would return to the things he knew best, leaving revolutionary politics to others.

It was at about this time that he began designing the bridge across the River Severn at Montford (described earlier), 4 miles west of Shrewsbury, which supported 'the great high road from Shrewsbury into Wales'. It was his first bridge in Shropshire, the first built anywhere under his superintendence, and surely what was needed to re-stabilise him after his obsessions with Dorothea Jordan and Tom Paine. Civil infrastructure executed 'in the service of man' would increasingly become Thomas Telford's hallmark, his life's calling, the practical expression of his special genius.

In the same year, 1792, that the Montford Bridge was completed, he was engaged to design and build a new parish church in Bridgnorth, Shropshire. His reputation was growing, and he was increasingly accepted as an architect rather than a constructor of other men's designs. According to Samuel Smiles the church of St Mary Magdalen stands

> at the end of Castle Street, near to the old ruined fortress perched upon the bold red sandstone bluff on which the upper part of the town is built. The situation of the church is very fine, and an extensive view of the beautiful vale of the Severn is obtained from it. Telford's design is by no means striking, being, as he said "a regular Tuscan elevation; the inside is as regularly Ionic: its only merit is simplicity and uniformity; it is surmounted by a Doric tower, which contains the bells and a clock". A graceful Gothic church would have been more appropriate to the situation, and a much finer object in the landscape; but Gothic was not then in fashion – only a mongrel mixture of many styles, without regard to either purity or gracefulness. The church, however, proved comfortable and commodious, and these were doubtless the points to which the architect paid most attention.

Whether or not a 'graceful Gothic church' would have been more appropriate is, of course, a matter of taste. Smiles certainly knew how to damn with faint praise; but he was writing many years later, at a time when the Gothic Revival movement was in full

St Mary Magdalen in Bridgnorth, designed by Thomas Telford and completed in 1795 (Geograph/James Allan; Wikipedia/Andrew Abbott).

Looking over the River Severn from Bridgnorth High Town (Geograph/James Allan).

flow in England, challenging neoclassical styles and harking back to the flamboyance of Christian medieval architecture. We should not be surprised that Telford's first adventure in church design drew heavily on his formative neoclassical experiences as a stonemason in Edinburgh and London. He was playing safe with what he knew, and many of us would agree with him; in any case the situation of his church is as fine today as ever, with extensive views over the delightful Vale of Severn spread far below.

His next step was a tour of major towns and cities in the south of England to widen his knowledge of the 'best forms' of architecture. He visited Gloucester and Worcester, and was enchanted by a detour through the manufacturing districts of Gloucestershire, especially the fine scenery of the Vale of Stroud which seemed to him 'a smiling scene of prosperous industry and middle-class comfort'. He spent several days in Bath before taking the coach to London, where he re-examined the principal public buildings with a newly expert eye; then returned to Shrewsbury via Oxford, which he adored, and Birmingham, which he loathed, judging it 'famous for its buttons and locks, its ignorance and barbarism'.

Back in Shropshire he resumed his work as county surveyor, overseeing roads and bridge repairs, and organising the convicts. The 35-year-old was still recognisably the affable 'Laughing Tam' of his youth, but his independent spirit and increasing confidence meant he was no pushover. Above all, his hard work and competence continued to impress influential figures in the county, and soon led to his next major, and entirely unexpected, move:

> An unforeseen circumstance, though not a hindrance, did very shortly occur, which launched Telford upon a new career, for which his unremitting study, as well as his carefully improved experience, eminently fitted him: we refer to his appointment as Engineer to the Ellesmere Canal Company.

As we shall see, the appointment was not exactly as principal engineer to the company, but rather as resident engineer and assistant to one of the most respected canal engineers of the age, William Jessop. Nevertheless it was an extraordinary vote of confidence in a man who had absolutely no experience of canal engineering, offering him a ladder that would lead, step by step, to national and international fame.

4 The Llangollen Canal

My favourite canal author is L.T.C. (Tom) Rolt, especially for his famous book, *Narrowboat*, first published in 1944, and for *Landscape with Canals* published in 1977, three years after his death. Tom Rolt had a special love of the many canals he had helped rescue from dereliction, and the Llangollen – a modern derivative of the Ellesmere Canal which Thomas Telford had been invited to help build in 1793 – was arguably his favourite. He dreamt for many years about coaxing his narrowboat, *Cressy*, across Telford's Pontcysyllte Aqueduct, an ambition that in due course infected me.

But the Llangollen was a long way from home. We kept our first narrowboat, a 48-foot cruiser, on the Kennet and Avon Canal, within easy reach of Bristol. We travelled the length of the K&A many times, venturing onto the Thames at Reading and, according to our mood, east to London, west to Gloucestershire, or north towards Birmingham via the Oxford and Grand Union Canals. Over 25 years my wife and I discovered areas of England we hardly knew existed, and were constantly delighted by the countryside crossed by great canal engineers of the late 18th and early 19th centuries as they developed England's transport infrastructure.

As far as boating challenges were concerned, perhaps I should have been satisfied with John Rennie's superb lock staircase at Caen Hill on the K&A, which we ascended and descended several times. But Pontcysyllte was still there, refusing to be ignored.

My wife and I were always snail-like boaters, preferring to cover a leisurely 10 or 12 miles a day rather than double that amount at maximum speed. Why rush around the canal system as though on land, when three quarters of its charm is the relaxing pace, and the lengthy chat in a canalside pub followed by exploration of the local village? Unfortunately 650 canal miles to the Llangollen and back at our preferred pace would have taken seven or eight weeks, time we could not afford. So, for the first and last time in our boating career, we drove post-haste to Shropshire and hired a narrowboat.

The Ellesmere Canal, Telford's introduction to canal engineering, was a huge undertaking planned for Shropshire, Cheshire and Wales in the heady days of England's canal mania. The history of its construction and ownership was exceptionally convoluted. The original plan had envisaged a strategic north–south connection from Liverpool to Bristol via industrial north-east Wales and the manufacturing towns of the West Midlands region. This meant cutting the Ellesmere Canal from the embryonic – and optimistically named – Ellesmere Port on the River Mersey, via Chester and Wrexham, passing near the small Shropshire town of Ellesmere, then down to the River Severn at Shrewsbury. Various branches would serve the needs of local landowners and industries.

When shares were put on offer in Ellesmere in September 1792 there was a stampede to invest. A consenting Act of Parliament was passed in 1793, and one of England's most experienced engineers, William Jessop, was appointed to take charge. But the Ellesmere Canal, as originally conceived, was never built. After endless commercial and political machinations by many interested parties the completed sections were eventually merged into the Shropshire Union canal system in 1846.

In 1944 much of the 'Shroppie' was closed by the then owner, the LMS Railway; but it retained the canal's main line and kept the branch to Llangollen open as a vital channel

A map showing the Welsh and Llangollen Branches of the Shropshire Union, now rebranded as the Llangollen Canal. Starting at Hurleston Junction on the main line of the Shropshire Union, it travels roughly southwest towards Ellesmere, then northwest to the Pontcysyllte Aqueduct and Llangollen.

The original plan for the Ellesmere Canal was from Ellesmere Port to Chester, to Wrexham and then south-westwards to Trevor; then over the Pontcysyllte and down the Weston Arm of the Montgomery Canal to Shrewsbury.

The Hurleston lock flight at the eastern end of the Llangollen Canal (Geograph/Roger Kidd).

supplying water from the upper reaches of the River Dee. As a result, today's Llangollen Canal comprises two sections, originally part of the Ellesmere Canal and subsequently known as the Welsh and Llangollen branches. Ironically, it carries more traffic than it ever did as a commercial waterway, as it is hugely popular with boaters and holidaymakers.

We will now begin a cruise along its 46-mile length, beginning at Hurleston Junction near Nantwich in Cheshire, crossing from England into Wales near the Pontcysyllte Aqueduct, and ending just beyond the town of Llangollen. This is the country that introduced Thomas Telford to the myriad challenges of canal engineering and gave him the opportunity to design and build one of the world's greatest aqueducts.

As we leave the main line of the Shropshire Union at Hurleston Junction there is pressing work to do: the ascent of four locks. A total rise of 34 feet starts to open up the surrounding countryside, except on our right where the long embankment of Hurleston reservoir temporarily obstructs the view. It is a fair bet that most boaters passing this way pay scant attention to the reservoir's 24 acres of placid water – unless they enjoy carp fishing. But its presence alerts the engineers among us to its role in topping up the Shroppie and supplying water to local communities.

The Llangollen is a very unusual navigation canal in that there is a significant current flow along its entire length. You may recall that the five ancient aqueducts described earlier

– Eupalinos in Greece, Petra in Jordan, Hampi in India, Segovia in Spain, and Nîmes in southern France – were all used for the transport of water, not for navigation; and that the word 'aqueduct' has often been used to describe the complete length of a water channel, not just its bridges. So I have a suggestion for lively canalside discussion: our 46-mile canal, used for water transport *and* navigation, might as well be renamed the 'Llangollen Aqueduct', and its most famous engineering structure 'Pontcysyllte Bridge'. But I am in danger of landing in water considerably deeper than Hurleston Reservoir!

The constant current flow means that our journey from Hurleston to Llangollen will take slightly longer than the return, because up to 14 million gallons per day, obtained from the River Dee near Llangollen, pass along the canal to maintain water levels and supply the reservoir. The flow is more noticeable in the narrower and shallower sections, including tunnels. Of course some water is released into the main Shroppie every time a boat passes through Hurleston locks, but the rest finds its way into the reservoir via a feeder at the top of the flight.

Passing under the busy A51 and A534 we say a grateful goodbye to road traffic and continue 6 miles through gentle Cheshire farmland. After negotiating two locks at Swanley and three at Baddiley, we come to Wrenbury Wharf, home to a hire-boat base, a former corn mill converted into an attractive pub, and a lift bridge complete with traffic lights.

Canal bridges that disrupt the flow of vehicles tend to try the patience of everyone concerned. I am reminded of an episode many years ago on the K&A near Newbury: eight in the morning, lashing rain, rush-hour traffic, commuters in cars impatient to get to their offices. My wife landed from our narrowboat with the appropriate key, and after waiting for a lull in the traffic tried to open the bridge. But it was extremely heavy, poorly maintained and, in those far-off days, un-motorised. It steadfastly refused to budge. We knew that boaters had right-of-way over car owners, so we persisted and eventually managed to move the wretched structure about half-way, where it stuck, as useless for boats as it was for cars. A miserable and embarrassing 20 minutes followed as vehicles built up on both sides of the bridge and started blowing their horns. Requests for help from the frustrated motorists

The raised embankment of Hurleston Reservoir, seen behind Lock No. 2; and the reservoir feeder above Lock No. 4 (Geograph/Roger Kidd).

The lift bridge and pub at Wrenbury (Geograph/Roger Kidd).

fell on deaf ears. We did eventually get the help we needed, in exchange for verbal abuse. Thankfully, motorisation of the bridge arrived a couple of years later.

No such problems at Wrenbury – the lift bridge is motorised and looks in good order as we approach; even better, a narrowboat in front of us has stopped the cars, raised the bridge, and is about to go through. We follow close behind, studiously avoiding eye contact with the motorists.

The next 6 miles, punctuated by four locks, lead us to one of the canal's best-known features – the six locks at Grindley Brook. The first three come in close succession; the next three are arranged as a staircase, which requires a certain amount of concentration.

In a 'real' staircase like Grindley Brook, the upper gate of one lock also acts as the lower gate of the lock above. It is probably best to think of the staircase as a single lock with intermediate levels. Operating it is very different from operating a normal flight, because a chamber can only be filled by emptying the one above, or emptied by filling the one below. This means that the whole staircase has to be full of water (except for the bottom chamber) before a boat starts to ascend; or empty (except for the top chamber) before a boat starts to descend.

In the variation known as an 'apparent' staircase, the water does not pass directly from one chamber to the next, but goes into substantial side ponds. The water level in a pond does not alter much when a single lockful is discharged into or taken from it, so it is not necessary to fill or empty the flight before starting. A famous example of an apparent staircase is Caen Hill on the K&A near Devizes, which consists of 16 close-spaced locks with side ponds.

The Grindley Brook locks are kept in excellent order these days, as they need to be. Even so, in the busy summer months they often create bottlenecks as narrowboats gather at both ends, and the lock keeper is kept extremely busy shepherding them through three at a time, either up or down. At times like these, it does us good to recall the experiences of author Tom Rolt, who tried but failed to coax his narrowboat *Cressy* along the semi-derelict Llangollen Canal towards Pontcysyllte in 1947, and tried again with more success in 1949.

Ascending the locks at Grindley Brook (Geograph/Roger Kidd).

Rolt's 1947 cruise was almost predestined to fail because no boats had voyaged from Hurleston Junction to Llangollen since before World War II, and in 1944 the LMS Railway had obtained Parliamentary permission to close the entire canal. He was permitted to enter it on sufferance, but was informed that he had no right of navigation and could expect no assistance if in trouble: it would be sink or swim.

On arrival at Hurleston Junction he and his wife found *Heron*, a small cabin cruiser owned by a Liverpool family who had also decided to venture along the Llangollen and see how far they could get. The two crews set off together, promising mutual support. Their first trial occurred at Baddiley, where the side walls of Lock No. 9, forced in by frost damage, captured *Cressy* halfway into the empty chamber in a vice-like grip. Nevertheless, they made it through. Beyond Baddiley they encountered severe water shortages and huge quantities of weed that fouled the propellers, and had to resort to 'bow-hauling'. Conditions were so bad that they considered a retreat to Hurleston, especially in view of a local rumour – false as it turned out – that Wrenbury Bridge was damaged and could not be lifted. But they pressed on and eventually reached Grindley Brook

> only to find that "Cressy" again stuck fast between the wing walls of the bottom lock. It was only after repeated flushing from the top paddles that eventually, to our great relief, she came free and we were able to tie up to the bank below the lock. Here we held a council of war. Somehow or other we had to get "Cressy" up the Grindley Brook flight, for the prospect of retreat did not now bear thinking about; it would have meant stern-hauling our heavy boat back to the "winding hole" [a short dead-end section, usually at right angles to the canal, that allows boats to turn round] in Wrenbury, and even then it was by

> no means certain that we could turn her. The day was Sunday, and we walked up to the lock house at the top of the flight where we explained our predicament to the lock keeper, Mr Howell, who promised to come down with an assistant to give us a hand next morning.
>
> I spent most of the rest of the day planing obvious high spots off "Cressy's" wooden hull. I also greased her bow rubbing strakes, for I had resolved upon the desperate expedient of driving at full speed towards the lock chamber. If she went in, well and good; if she did not, I realised that nothing would free her short of partly demolishing a lock wall as Brunel and Claxton had had to do in order to get the "Great Britain" out of Bristol Docks.

Not surprisingly, Mr Howell took some convincing that the plan was feasible. But with his agreement *Cressy* was taken 50 yards astern, put full ahead, and aimed towards the empty lock like a projectile. The plan worked perfectly. So did passage up the triple staircase. With huge relief, the Rolts moored opposite the lock keeper's cottage at the top of the flight, and went off to do some shopping in nearby Whitchurch.

The episode was just one example of the extraordinary determination and courage shown by Tom Rolt as he campaigned to keep the failing English canal system alive. He is surely the person above all others to whom we owe a huge debt of gratitude for today's narrowboating pleasures.

The charming bow-fronted lock keeper's cottage at Grindley Brook deserves our attention, because it was designed by Thomas Telford. If it looks more substantial than a cottage, this is because it also served as a toll house for collecting dues from passing narrowboats. It is easy to imagine William Jessop and the Ellesmere Canal Company applauding the new engineer on their staff who could turn his hand to designing office buildings. And if the result lacks the neoclassical influences of Somerset House or his church at Bridgnorth, it is because he decided to adopt an unfussy style more suited to modern canal infrastructure, complete with angled windows that gave the lock keeper a good lookout over the comings and goings below.

After leaving Grindley Brook the canal heads towards the delightful town of Whitchurch. Actually it was initially bypassed, and only connected by a short arm a few years later. Today even that arm shows signs of withering, and the centre of Whitchurch finds itself marooned a mile or so from the nearest canal water.

Two miles further on we begin to skirt Whixall Moss, a strange area of oozy peat bogs and small lakes, or meres, which forms part of the Fenn's, Whixall and Bettisfield Mosses National Nature Reserve. A Site of Special Scientific Interest (SSSI), it is one of the biggest and best raised bogs in Britain, with an astonishing variety of wildlife and bog plants, including the round-leaved insect-eating sundew.

The Llangollen Canal is also well known for its lift bridges, often light, well-balanced structures operated by human muscle power, and there are several in the Whixall area. The main thing to remember is that when raised they may not reach the vertical, but can

The lock keeper's cottage and toll house at the top of the Grindley Brook flight, designed by Thomas Telford (Geograph/Jo Turner).

A lift bridge near Whixall Moss (Geograph/Roger Kidd).

Prees Junction, looking back eastwards towards Hurleston and southwards towards Prees on the right. The 'roving bridge' on the left allowed horses to transfer sides without unhitching from their boats. (Geograph/Roger Kidd).

overhang the canal at an angle. Care is needed to avoid scraping the topside of a narrowboat on the way through – and remember that it is always better to 'duck, not grouse'!

We soon encounter a junction with another short arm of canal known as the Prees Branch. The original intention was to reach the village of Prees, but it only progressed as far as a group of lime kilns at Quina Brook. Coal was brought in to fire the kilns, and lime taken away for agriculture and industry. Today the truncated channel ends in a marina, the site of an old clay pit from which puddle clay had been extracted to line the canal.

Back in the 1790s William Jessop and Thomas Telford were less than delighted by the boggy ground of Whixall Moss, because they had to cross it. Fortunately Jessop had considerable experience of land drainage, learned among the fens of East Anglia and the Somerset Levels, and could advise his inexperienced colleague. They decided to make a straight one-mile cut across the western end of the Moss and, as the land dried out, build an embankment to take the navigable channel. It was a monumental task for labourers with little more than shovels and wheelbarrows, and would prove a valuable warning to George Stephenson as he struggled to cross the infamous Chat Moss during construction of the Liverpool and Manchester Railway 30 years later. Peat extraction from Whixall Moss has caused subsidence over the years, and up to the 1960s British Waterways employed a special Moss Gang to maintain the canal banks.

Shortly after clearing Whixall Moss the canal springs another surprise, an area dominated by water of quite a different character. The Shropshire Meres – sometimes, rather misleadingly, called the Shropshire Lake District – comprises many tree-fringed meres of shallow peaty water, whose basins had been formed in the underlying sandstone during the last Ice Age . The whole area is famous for plants and wildlife, including nearly 2,000 species of insect. We pass very close to two large meres, Colemere and Blake Mere, tree-surrounded, and secluded, favoured haunts of coot and moorhen. With luck, and like Tom Rolt and his wife before us, we can find a mooring alongside Colemere and enjoy a 'protracted picnic lunch'.

The 1-mile straight cut across the western end of Whixall Moss (Geograph/Christine Johnstone).

Next comes our first underground experience since leaving Hurleston: the Ellesmere Tunnel, completed by Thomas Telford in 1802. Its 87 yards are modest by canal standards, and minimal compared to the 2,900-yard Harecastle that he built on the Trent and Mersey 25 years later; but tunnels, like embankments and locks, were strangers at the start of his canal career, and he was probably grateful to be landed with a short one.

I admit that our own first underground experience, the 502-yard Bruce Tunnel on the summit pound (a pound is a length of canal between two locks) of the K&A near Crofton in Wiltshire, had been a bit daunting. We had heard stories from experienced boaters about getting stuck in a tunnel, or meeting a narrowboat coming the other way, or bouncing from side to side against the tunnel walls, or becoming disoriented in the dark. The Bruce was bored comparatively wide to take broad-beam local barges, but without a towpath, so in those early days the boatmen had to haul their boats through by pulling on chains attached to the inside walls. We had entered without much confidence, but all went well on the day – as, hopefully, would our passage through the Ellesmere, which was bored for a single narrowboat plus towpath.

Soon after leaving the tunnel we reach Ellesmere, the small town after which the Ellesmere Canal, and its planned terminal at Ellesmere Port on the Mersey, were named. It was here that investors, desperate to get aboard the canal mania bandwagon, were offered shares in the new company in 1792. As the main line of the canal curves sharply southwest

The Ellesmere Tunnel (Geograph/Eirian Evans), and its western portal (Geograph/John Haynes).

we skirt round Beech House, the company's impressive headquarters designed by Thomas Telford to provide a committee room, plans store, accounts office, and apartments for resident engineer, agent and accountant. Telford spent a lot of professional time here, designing, planning and directing, and we may imagine him looking out of the bow windows, up and down the canal's main line and, directly opposite, along the short arm that leads to Ellesmere Basin and the town wharf. The Ellesmere Arm was very much the centre of operations, with its warehousing, cranage, dry dock, and workshops.

We are now about halfway through our cruise, a good moment to take stock. The accompanying map shows the line of the canal from Ellesmere towards Chirk, Trevor and Llangollen. So far, our general direction of travel from Hurleston Junction has been southwest, but from now on it will be northwest towards Wales, with Pontcysyllte firmly in our sights.

Beech House, designed by Telford as headquarters of the Ellesmere Canal Company (Geograph/Jeremy Bolwell); and an old warehouse at Ellesmere Basin, with signage of the Shropshire Union Railways and Canal Company, subsequent owner of the canal (Geograph/Chris Andrews).

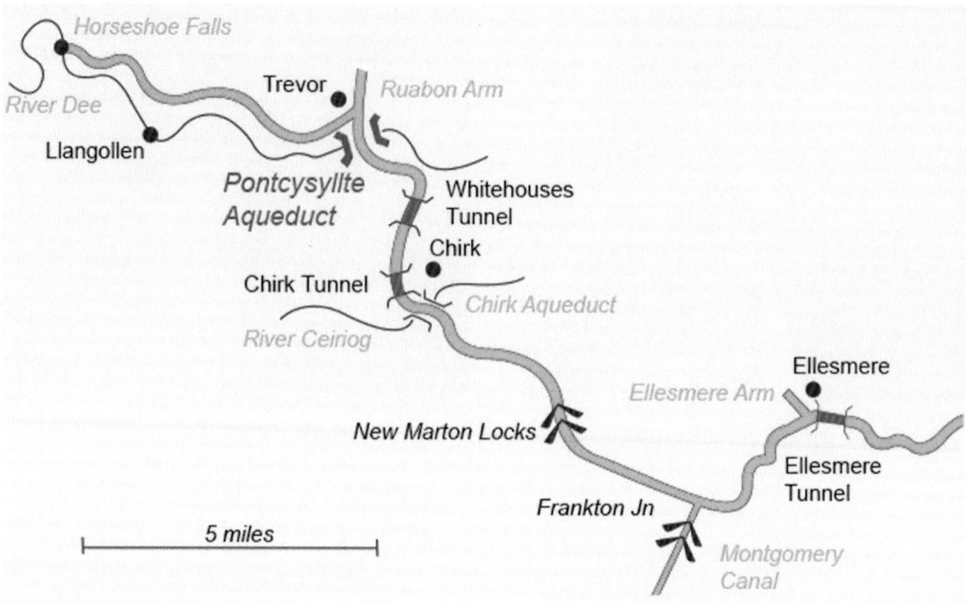

The canal from Ellesmere to Llangollen.

This was also the hope of Tom Rolt and his wife in 1947 as they pondered the state of the canal beyond Ellesmere. They had already bow-hauled the *Cressy* through a thick blanket of weed from Ellesmere Tunnel to Ellesmere Basin, and were wise enough to establish what lay ahead before passing such a convenient turning point. It could scarcely have been more discouraging: a canal choked by a veritable Sargasso Sea of weed as far as the eye could see; and even if they battled on they would get no further than Chirk Tunnel because of a landslip. Their ambition to cross Pontcysyllte scuppered, they retreated disconsolate to Hurleston Junction and the Shroppie. Once again, we must be thankful for the almost miraculous restoration of the English canals in the last half-century, and the calm confidence with which we can leave Ellesmere Basin and continue our own adventure.

The next talking point, unspectacular on the ground but pivotal in the history of the Ellesmere Canal, is Frankton Junction. This is where today's Llangollen Canal meets the Montgomery Canal, a waterway that has been undergoing valiant restoration, section by section, for many years.

If you were to head south along the Montgomery you would soon meet a double lock, two single locks and a T-junction. Turn right to enter a long stretch of canal towards Llanymynech and central Wales; turn left into the Weston Arm and, very soon, a dead end. And therein lies a story.

The Weston Arm had originally been planned as part of the Ellesmere Canal's main line, from Frankton Junction southwards to the River Severn at Shrewsbury, a distance of about 15 miles. But it never happened. The cut only progressed as far as the village of Weston Lullingfields, about 9 miles short of Shrewsbury, where the canal company built a wharf, four lime kilns, a public house, stables, a clerk's house and a weighing machine.

Frankton Junction and its signpost (Geograph/Roger Kidd).

These were opened in 1797 and were closed in 1917 following a breach of the canal. The Weston Arm, so insignificant today, had originally been part of William Jessop's dream.

Nowadays the junction at Frankton is largely ignored by a boating fraternity keen to press on towards Pontcysyllte and Llangollen. Certainly it looks innocuous enough in its tranquil setting, and few of us give the 90-degree turn into the Montgomery Canal much thought. But we might pause to imagine William Jessop here in 1792, perhaps in the very spot where a signpost now stands, surveying a route for the main line of the Ellesmere Canal. Frankton Junction would be roughly its centre point, from where it would run towards Shrewsbury and the River Severn in one direction, and to Pontcysyllte Aqueduct, Trevor, Wrexham and Ellesmere Port in the other. He could hardly have predicted that rising construction costs and a failure to generate enough commercial traffic would put paid to the grandiose scheme, or that the parts which were completed would become sections of the Shropshire Union System. In fact most major building work on the Ellesmere Canal ceased following the completion of Pontcysyllte in 1805. The main line was never completed, leaving the Ellesmere with a good claim to be 'the canal that never was'.

The historic muddle surrounding the Ellesmere Canal's development has resulted in a curious numbering system for the bridges on today's Llangollen. From Hurleston Junction to Frankton Junction the numbers go from 1 to 70, and continue along the Montgomery Canal. But west towards Llangollen they start again at 1, with suffix W. Thus Rowson's Bridge at Frankton, No. 70, is also known as No. 1W. All this discussion of canal routes and bridge numbers may have obscured the fact that we have not negotiated a single lock since leaving Grindley Brook. The unusually long pound reflects the flat landscapes of Whixall Moss and the Shropshire Meres. But as we continue our cruise beyond Frankton, the

New Marton Upper Lock (Geograph/Roger Kidd).

country becomes slightly more undulating and we encounter the first locks for 20 miles, delightfully situated near the hamlet of New Marton. What few boaters realise is that by the time they have negotiated the upper lock they are floating on an uninterrupted ribbon of water that includes Telford's 'stream in the sky' at Pontcysyllte and the vital feeder from the River Dee near Llangollen.

The next few miles seem a little dull by comparison. Although the countryside is predominantly agricultural, the large village of St Martins, off to our right, gives notice that we are entering a former mining area. The colliery here was the largest in Shropshire, and its workings crossed the border into Wales, confirming it as a geological extension of the Denbighshire coalfield. It closed in 1968. A little further on, Chirk Bank Colliery was opened in 1801 and closed in the 1880s. It seems to have triggered a collapse of the canal embankment in 1816, obstructing the River Ceiriog downstream and flooding a number of pits in the valley. Human lives were spared, but the pit ponies perished. Today, Shropshire seems a quiet agricultural county; but it was not always so.

Approaching Chirk Bank the canal springs another surprise. After a brief dalliance with Telford's famous Holyhead Road, the A5, it swings west to follow the contour of a steep-sided valley, maintaining height and giving a fine view over the River Ceiriog. We also get an occasional glimpse of Chirk Aqueduct, our next adventure. The following 11 miles of hill country promise to reveal more about Telford's engineering genius than the 35 we have covered since leaving Hurleston Junction.

The Ceiriog is a tributary of the River Dee. Rising at an altitude of about 1,800 feet in the Berwyn Mountains, some 12 miles west of Chirk, its steep gradient and fast flow give it a reputation as one of the wildest rivers in Wales. After skirting Chirk Castle it joins the Dee east of the town. The river and its beautiful valley were described by David Lloyd George, Prime Minister between 1916 and 1922, as 'a little bit of heaven on earth'.

The River Ceiriog and Chirk Aqueduct, seen from Chirk Bank (Wikipedia/Cls14).

Jessop had some hard decisions to make about crossing the Ceiriog and the Dee. Both river beds are far lower than the canal at Chirk Bank, which, as our muscles remind us, has climbed from Hurleston Junction via 21 locks – a total rise of about 150 feet. Like the aqueduct engineers of ancient Rome and the English canal pioneer James Brindley, Jessop knew there was much to be said for preserving precious height by following the contour. But two valleys stood in the way, and would require enormous bridge aqueducts. There was also a summit problem. All canals with locks have one or more summit levels, and successful operation demands that those summits are well supplied with water. Every boat going over a summit releases two lockfuls to the pounds below – about 50,000 gallons for narrow locks like those on the Llangollen – and the water must be replaced swiftly and efficiently.

After going over the ground himself in the early summer of 1792, William Jessop recommended a line which by no means met these criteria. Even so it would have made the Ellesmere one of the costliest canals ever built in Britain. Dropping from Chirk Bank with a flight of locks to the valley below, he would cross the Ceiriog with a low-level aqueduct; bore a mile-long tunnel through to the Dee valley; cross the Dee with another aqueduct; then climb up to Trevor using more locks. This would create summit levels at Chirk Bank and Trevor, and a long pound across the two valleys that would act as a sump – the exact opposite of a useful water supply. Indeed Jessop would go further and bore a formidable 4,600-yard tunnel beyond Trevor to push the canal's northern summit on towards Wrexham.

Even his most enthusiastic supporters were daunted by the 'Great Tunnel', as they called it. The two summits would need plentiful water supplies, but their sources were unclear. Descent into the Ceiriog valley and ascent out of the Dee valley would require extensive lockage. Unsurprisingly, Jessop was asked to return to the drawing board and reconsider.

As luck would have it, Thomas Telford joined William Jessop in 1793 before the matter was finalised and, in particular, before a final decision was reached about the two aqueducts. Should they be low-level and relatively straightforward, or could they be elevated to maintain the contour at Chirk Bank right through to Trevor? To most civil engineers of the time, the answer would have seemed obvious: a sky-high structure over the River Dee would be asking for trouble. Nothing comparable had been attempted in the Western world since the Romans had completed the Segovia Aqueduct and the Pont du Gard. And surely no one would seriously suggest a monumental tiered aqueduct of stone, topped by a navigable conduit puddled with clay – as risky for the engineer's reputation as it would be for the coffers of the Ellesmere Canal Company. Fortunately, Telford had become fascinated by an alternative engineering material – cast iron – and had the vision and daring to champion its use.

Nothing could better illustrate the company's extravagant ambition than its final decision to build two high-level aqueducts and two tunnels to maintain the water level reached at Chirk Bank. The costs were enormous, the technical challenges huge, the risks daunting. Yet the company decided not only to build the new canal through difficult country, joining the mining and iron industries of north-east Wales to the growing canal network, but to provide a fast through route with minimal lockage from Liverpool to Shrewsbury and

A narrowboat chugs across Telford's Chirk Aqueduct while a train speeds over the parallel railway viaduct (Wikipedia/Akke Monasso).

Chirk Aqueduct and railway viaduct seen from the Ceiriog Valley (Wikipedia/Akke Monasso).

the Severn. The short-lived but frenzied canal mania of the early 1790s ensured that the scheme was enthusiastically received and heavily oversubscribed.

So it is worth recalling that we are now cruising one of the most expensive, challenging and extraordinary canal pounds in Britain. And as our boat swings gently north to cross the River Ceiriog we come face to face with the first of Telford's great structures, the 10-span, 70-foot-high Chirk Aqueduct. Its fame would be even greater were it not for neighbouring Pontcysyllte. Look west under the arches of the parallel railway viaduct built half a century later, enjoy the intense green of the valley floor beneath the hanging woods of 700-year-old Chirk Castle, and celebrate our arrival in Wales by one of the most dramatic routes imaginable. Not surprisingly, the aqueduct also marks the start of an 11-mile stretch of World Heritage canal.

But don't get too carried away, especially if you are steering, because Chirk Aqueduct is followed almost at once by Chirk Tunnel, the second major engineering work built to preserve the canal's level. As we might expect on entering Wales, the hills on this side of the valley are higher than those on the English side. The silver ribbon of aqueduct water gives way to a short canal basin, and then almost immediately to a leaden ribbon underground.

The entrance to Chirk Tunnel and also, as the signpost proclaims, into Wales (Geograph/ Roger Kidd).

Inside Chirk Tunnel (Geograph/Roger Kidd).

Just over a quarter of a mile long, the one-way Chirk Tunnel was completed in November 1801. Thomas Telford thought James Brindley's practice of 'legging' boats through canal tunnels degraded people to mere beasts of burden and would have nothing to do with it. So he gave his tunnel one of the first-ever towpaths for horses. Luckily it is supported on brick arches that allow water to flow beneath and reduce the drag on passing boats. Even so, we should prepare for a noticeable drop in speed as we enter the southern portal, remembering that the canal is constantly feeding water down to Hurleston Reservoir. The tunnel is dead straight, so we can check that there is no competition coming from the far end. It is also brick-lined and comfortably dry thanks to its 'cut and cover' construction – Telford placed a seal of clay on top of the brickwork vault before backfilling with loose stones and excavated spoil.

Coming out of Chirk Tunnel at the north portal (Geograph/Roger Kidd).

We emerge from Chirk Tunnel into a long tree-shaded cutting, followed by a winding hole, the site of a former transhipment wharf. The output of mines and quarries further up the Ceiriog valley was trundled to the wharf by the narrow-gauge Glyn Valley Tramway for onward transport by narrowboat. This, too, was a site of intense 19th-century industrial activity.

Soon afterwards we pass through Telford's 174-yard Whitehouse Tunnel, also built using cut and cover and furnished with a towpath, which was opened for navigation in 1802. He must have enjoyed routing his Holyhead Road over the top 20 years later.

Two straight miles beyond Chirk we pass under Irish Bridge, No. 27W, named after the huge number of Irish navvies who tramped across North Wales from Holyhead to help build the navigation. The canal now swings west, very much as it did on the approach to Chirk, and for the same reason – to preserve precious height by following the contour of a valley side – not, this time, of the Ceiriog, but of the Dee, a much more substantial affair. Rising in Snowdonia, the Dee has a larger catchment area, a much greater average flow, and by the time it reaches the Vale of Llangollen and Pontcysyllte, a much wider valley.

Whitehouse Tunnel (Geograph/Roger Kidd).

Irish Bridge, No. 27W (Geograph/Roger Kidd).

A glimpse of Pontcysyllte from the southeast (© Crown copyright: RCAHMW).

With just a mile and a half to go, it is hardly surprising if pulses start to race a little. Tom Rolt and his wife felt the same in 1949 as their renewed attempt to reach Pontcysyllte approached its climax. They had found the canal from Hurleston Junction in considerably better shape than two years previously – no trouble at Grindley Brook locks, no Sargasso Sea of weed at Ellesmere, no rumour of a landslip near Chirk. As they swung the *Cressy* westwards along the Vale of Llangollen they had their first glimpse of the incredibly tall, elegant stone piers that promised to carry them over the River Dee. But instead of hurrying onwards, they tied up for the night at Froncysyllte Wharf, the last before the aqueduct, to await the arrival of friends for the big event next morning.

The large village of Froncysyllte, colloquially known as Fron, is situated on Telford's London to Holyhead road as well as on his canal (the anglicised spelling Vroncysyllte was commonly used until the mid-20th century, including by Tom Rolt). The name is derived from the Welsh for a hillside, together with Cysyllte, one of the old townships of Llangollen Parish. Its cottages were built for quarry, limekiln, brick and tile workers during the 19th century and although the area's traditional industries have now gone, a rich industrial archaeology remains.

Froncysyllte is also known for its male voice choir, formed to coincide with the inaugural Llangollen International Musical Eisteddfod in 1947, just two years before the Rolts' successful cruise. Young men were encouraged to make up the minimum of 60 voices needed to compete in the Eisteddfod. Eight years later they hosted a male voice choir from Modena in Italy, which won first prize at the Eisteddfod. One of its members was 19-year-old tenor Luciano Pavarotti, who considered it the most formative experience of his life, the inspiration needed to become a professional singer.

Somehow the combination of industry and music-making in Froncysyllte seems particularly suited to a village whose near neighbour, Pontcysyllte, is a masterful blend of engineering and art in its widest sense. Tom Rolt, an engineer of great artistic

Approaching Froncysyllte (Geograph/Roger Kidd).

sensitivity, must surely have felt the same as he pondered the next morning's long-anticipated cruise.

Entry onto the aqueduct is sudden. After all the miles, and locks, and meres, and tunnels, we round a tree-lined bend, straighten up for 100 yards or so, and notice the canal narrowing for a magical transfer from solid land to something altogether more ethereal, an impossibly long and slender ribbon of silver: Telford's magical 'stream in the sky'.

The solid land is itself quite remarkable at this point because it was formed by hundreds of navvies using spoil excavated from the Chirk and Whitehouse tunnels and the cutting between them. Telford could not pass almost trouble-free from land to aqueduct as he had at Chirk; he had first to engineer the approach. Tom Rolt described the aqueduct's towering south embankment as a unique feat of civil engineering, standing no less than 97 feet above the valley at its far tip and, with the possible exception of prehistoric Silbury Hill in Wiltshire, the greatest earthwork that had ever been raised in Britain at that time.

Emotions are hard to express as we launch out along the aqueduct. I prefer to avoid writer's responsibility by quoting Tom Rolt, whose determination was finally rewarded by a fantastic view over the valley 126 feet below:

> This panorama, thus suddenly disclosed, was breathtaking, particularly when combined with the slow, gliding motion of a boat which one automatically associates with lush and level water-meadows, willow trees and browsing cattle. No wonder the romantic writers and artists of the early nineteenth century were united in their praise of the engineer:
>
> TELFORD, who o'er the vale of Cambrian Dee
> Aloft in air at giddy height upborne
> Carried his navigable road …
>
> wrote Robert Southey. For engineering was then rightly considered by artist and engineer alike as an art rather than a science, and Pont

The challenge considered, and accepted (Paul A. Lynn).

Pontcysyllte from the air. Part of the southern approach embankment is on the left (© Crown copyright: RCAHMW).

Cysyllte aqueduct provided precisely that dramatic contrast between the art of man and natural beauty that they most admired, using such adjectives as "awful" or "sublime". As, with *Cressy*'s engine slowed to a tick-over, we crept out on to the aqueduct, I found myself in complete sympathy with their ardour.

Stream in the Sky (Geograph/John M. Wheatley).

Sky above, Earth below (Wikipedia/Akke Monasso).

Nineteen arches and just over 1,000 feet later we find ourselves supported once more by terra firma. The canal bends sharp left almost immediately, hugging the north side of the valley towards Llangollen; but before we follow it, a few words about the basin immediately opposite the far end of the aqueduct.

Trevor Canal Basin was originally built as the next stage of the Ellesmere Canal's main line, from where it would run north through heavily industrialised Cefn Mawr, climb by locks to a summit level near Ruabon, and service the Wrexham coal and iron district before proceeding to Chester and Ellesmere Port. So a narrowboat skipper, having crossed the aqueduct, would have continued straight ahead towards Wrexham rather than turning left towards Llangollen. But as we have seen, that stage of the main line was never built, and the canal was deprived of its intended summit water supply: Trevor Basin was left as the northern terminus of a sadly diminished Ellesmere Canal.

Today the basin is home to narrowboats and local businesses. Boats may be hired; historical artefacts recalling the aqueduct's construction explored in a visitor centre;

Shapes and shadows (Geograph/Chris Gunns).

Pontcysyllte and Trevor Basin from the air, with the southern approach embankment at the top and, centre right, the navigable feeder heading westwards towards Llangollen (© Crown copyright: RCAHMW).

and refreshment taken at the Telford Inn, initially the residence of Matthew Davidson who was chosen by Telford to supervise construction of the aqueduct. Built by the Ellesmere Canal Company in 1792, the building was originally known as Scotch Hall. It remained in private hands until 1981, then changed to a pub.

Tom Rolt and his wife decided not to attempt the final few miles to Llangollen. The navigable feeder is relatively narrow and shallow, and in 1949 its effective width had been further reduced by large amounts of clay puddle heaped against the bank to prevent water seepage and collapse – as had happened a few years earlier when a huge cascade of water and rubble caused derailment of a train passing along the Ruabon–Llangollen railway below. And even if the Rolts struggled on to Llangollen, the prospects for turning a 70-foot narrowboat around at the far end were poor, and the thought of stern-hauling *Cressy* the 4 miles back to the aqueduct daunting. So they moored about 200 yards beyond Pontcysyllte and stayed there for the best part of three months 'in perfect summer weather'. At one point *Cressy* met an old friend, an elderly craftsman who had helped build her in a dry dock in Trevor Basin during World War I.

The final section of canal from Pontcysyllte to Llangollen is daringly terraced along the valley side. Its design as a navigable feeder, and the difficult terrain through which it was cut, make it quite a challenge for today's narrowboaters. I well remember an acquaintance, more experienced than we were, warning that we were likely to spend our time 'bumping from side to side and scraping along the bottom'. He even congratulated us for hiring a narrowboat, rather than subjecting our own to an unwelcome dose of wear and tear. It turned out less awkward than that, but still needed caution, especially

Trevor Basin and the Telford Inn (Geograph/Philip Halling).

where the steepness of the hillside meant that the canal had been cut just wide enough for a single boat. And, as expected, the slight but steady flow of water against our bow made outward progress noticeably slower than the return.

Whatever difficulties the navigable feeder presents, the last few miles along one of the most beautiful stretches of canal in Britain are hugely enjoyable. There are some fine fixed bridges and on the approach to Llangollen, delightful houses and gardens. The Old Wharf in Llangollen, restored to provide a canal museum and gift shop, overlooks the town and acts as a base for horse-drawn tourist boats. A winding hole beyond the wharf marks our own limit of navigation.

But this is not the end of the navigable feeder. The River Dee is still well below us at this point, and we must follow the valley upstream until it meets the canal's summit level. This occurs about 2 miles beyond Llangollen, at a village called Llantysilio. Here, Telford decided to build a 460-foot curved weir, known as Horseshoe Falls, across the river, creating a large pond and continuous feed to the canal via a valve house and flow meter. The weir and navigable feeder were authorised by an 1804 Act of Parliament, and completed in 1808.

The weir was a major factor influencing the directors of the LMS Railway, owner of the Shropshire Union Canal, when they decided to close so much of the system in 1944. As the River Dee was a vital source of water for the main line of the Shroppie, they realised that the feeder had to be preserved, right down to Hurleston Junction. The millions of

Millars Bridge, No. 35W, over the navigable feeder (Geograph/Roger Kidd).

A narrowboat approaches Llangollen (Geograph/Roger Kidd).

The Old Wharf at Llangollen (Geograph/Richard Croft).

The River Dee passes vigorously through Llangollen town (Geograph/Steve Daniels).

Horseshoe Falls at Llantysilio, the weir built by Thomas Telford (Geograph/Ian Capper).

gallons that flow from Llantysilio each day soon find themselves traversing two majestic aqueducts, three tunnels, 21 locks, and landscapes as varied as the valley of the Ceiriog and Whixall Moss. And fortunately for us, they also support narrowboats.

5 *Pontcysyllte*

Great preparations

> I was last Monday appointed sole agent, architect, and engineer to the canal which is projected to join the Mersey, the Dee, and the Severn. It is the greatest work, I believe, now in hand in this kingdom, and will not be completed for many years to come … I had no idea of any such appointment until an application was made to me by some of the leading gentlemen, and I was appointed, though many others had made much interest for the place. This will be a great and laborious undertaking, but the line which it opens is vast and noble.

Vast and noble it certainly was. And, some would say, crazy. Thomas Telford's letter to his Langholm friend Andrew Little reveals excitement at being catapulted into the canal mania enveloping England, and the speculative frenzy that accompanied it.

Successful canals were already making their investors large amounts of money, and as the mania developed previously sceptical landowners became ever more anxious to join the party. The first public meeting of would-be investors in the Ellesmere Canal was packed with enthusiasts who were well aware that the Duke of Bridgewater's pioneering canal near Manchester, opened 30 years before, had brought him immense wealth. In anticipation of mind-boggling dividends they subscribed four times the projected cost – even though the navigation would pass through some extremely difficult country.

We should remember that Telford had been working as Surveyor to the County of Shropshire for six years and had advanced his reputation from that of excellent stonemason, via competent builder, to budding architect. He was known to be extremely hard-working and well-organised, but when the call came in 1793 he had absolutely no experience of canal engineering. His claim that he had been appointed 'sole agent, architect, and engineer' therefore seems a little overblown.

In fact, as previously noted, William Jessop had already been engaged to oversee the building of the Ellesmere Canal. If anyone deserved the title Engineer with a capital E, it was Jessop. What the company now needed was a highly competent engineer to take day-to-day charge of the project, with Jessop available to make major decisions and give advice and direction. Telford's appointment was, in effect, as 'resident engineer', with added responsibilities as 'sole agent and architect' – a highly demanding job description, but not quite the complete package. Little can he have imagined, at this early date, that it would give him the chance to design and build one of the world's most spectacular aqueducts.

Not that he had initially dreamt of offering himself for the job. Instead he was approached by some of the county's 'leading gentlemen', already admirers; and supported, without his knowledge, by the Shropshire ironmaster William Reynolds. But there were other candidates, including engineers John Duncombe and William Turner, who had been involved in the project from the start and might be expected to resent a relative tyro. Turner enjoyed Jessop's support, but Telford prevailed. In part this was because he was finally backed by John 'Iron-Mad' Wilkinson, the pioneering industrialist who had championed building the world's first cast-iron bridge, across the River Severn at Coalbrookdale. Wilkinson was heavily involved in iron production in the Black Country, Shropshire, and from 1792 onwards north-east Wales.

The controversial appointment was confirmed by a meeting of the proprietors held in Ellesmere at the end of October 1793. It was a rancorous occasion. Some doubted the wisdom of Jessop's recommended main-line route, others gave vent to professional and personal rivalries. As Telford put it:

> Besides the real labour that attends such a great public work, contentions, jealousies, and prejudices are stationed like gloomy sentinels from one extremity of the line to the other. But, as I have heard my mother say that an honest man might look the Devil in the face without being afraid, we must just trudge on in the old way.

Over the years there has been continuing debate about Telford's precise contribution to the Ellesmere Canal. This is partly because contemporary documents often referred to him as 'General Agent', whereas Jessop was described as 'Principal Engineer' and Duncombe and Turner as 'Engineers'. But the exact terms of Telford's appointment make clear the scope of his responsibilities:

> To attend meetings, make reports, to superintend the cutting, forming and making of the Canal and taking up and seeing to the due observance of the Levels thereof, to make the Drawings and to submit such drawings to the consideration and correction of Mr Wm. Jessop or the person employed by the said Company for the time being as their Principal Engineer. To give instructions for contracts ... to pay Contractors and workmen and to keep accounts. His engagement to extent to all Architectural and Engineering business, to the drawing, forming and directing the making of bridges, aqueducts, tunnels, locks, buildings, reservoirs, wharfs and other works.

Although several others deserve credit, Telford was surely the person who made decisive design contributions to the Ellesmere Canal and its modern derivative, the Llangollen. If one man's involvement has been unfairly downplayed, it is that of Jessop, an outstanding engineer whose name is still hardly known to the general public.

William Jessop (1745–1814) was a Devonshire boy, son of a foreman shipwright in Devonport Naval Dockyard. He was offered an apprenticeship by John Smeaton, Britain's first self-proclaimed 'civil engineer', a famous designer of canals, bridges, harbours and lighthouses who introduced the young man to canal engineering on the Aire and Calder Navigation in Yorkshire. Jessop's first major work as an engineer in his own right was on the Grand Canal of Ireland, linking Dublin with the River Shannon, which had begun as a government project in 1753 but which had taken 17 years to progress a modest 14 miles from the Dublin end. In 1772 a private company took over the work and consulted Smeaton, who sent Jessop across the Irish Sea as resident engineer. Jessop re-surveyed the proposed line of the canal; identified sources of water and built reservoirs; carried the canal over the River Liffey on the Leinster Aqueduct; and drove it across the notorious Bog of Allen on a high embankment.

Having seen to all the important details, Jessop returned to England, leaving a deputy in charge, and began to attract other important work. In 1789 he was appointed Engineer to the 14-mile Cromford Canal in Derbyshire, involving an aqueduct, several tunnels and 14 locks. In 1790 he helped found the Butterley Iron Works in Derbyshire, to manufacture (amongst other things) cast-iron edge rails. By 1792–3 his reputation was skyrocketing, and he became involved in several new canals including, above all, the Grand Junction, a hugely demanding project designed to link the Midlands directly to London, to which he was appointed Engineer.

So by early 1793 Jessop had an enormous amount on his plate. It is hard to imagine him adding the Ellesmere Canal, and unsurprising that having done so, he needed a highly competent resident engineer to design the details and oversee construction. His preferred candidate was William Turner, but when Telford got the job he accepted it with good grace and immediately started to work amicably with a colleague 12 years his junior. Jessop was never jealous of rising talent, but encouraged it – a sign of his modesty, good nature and reluctance to blow his own trumpet. These admirable but unusual qualities have left him inadequately acknowledged for his many remarkable achievements.

William Jessop (Wikipedia); and his Leinster Aqueduct over the River Liffey (Wikipedia/ Chris55).

Thomas Telford must have been reassured by the presence of the eminent engineer, whose Irish experience included building reservoirs, carrying a canal over a river, and driving it across a notorious bog – all highly relevant to the challenges posed by the Ellesmere Canal. He was fortunate to have such expert backing.

The 'leading gentlemen' agreed that Telford could continue his private architectural work. He was also determined to keep in touch with his friends, William Pulteney and Lady Bath, with whom he continued 'on intimate and friendly terms'. But he would give up the county surveyorship, and minor sources of income involving 'a great deal of very unpleasant labour for very little profit; in short they are like the calls of a country surgeon'.

Even so the scope of his new responsibilities meant that from now on he was heavily engaged with the Ellesmere Canal Company. The salary of £300 per annum was hardly excessive, but the honour and challenge of the appointment meant far more to him than money. His attitude was basically unchanged from the stonemason days at Somerset House in London: money was a means to an end, not an end in itself. And the end in his case was, increasingly, to become an outstanding civil engineer.

The Act of Parliament authorising construction of the canal was passed in 1793, and when Telford started work in October of that year he found himself in a team that included John Duncombe and William Turner as assistant engineers – hardly a happy situation. It seems that Duncombe accepted his subordinate role philosophically, but Turner was far more awkward. Fortunately, team tensions were offset, for Telford at least, when he secured the appointment of a schoolmate and good friend to the key post of Inspector of Works.

Matthew Davidson (1755–1819) was a native of Eskdale and a fellow stonemason from the Langholm days, who had been helping Telford build the Montford Bridge across the River Severn. The two men, now in their late thirties, trusted each other without question and would remain colleagues for the rest of their working lives. Telford regarded Davidson as the senior superintendent on the canal, and recommended Scotch House at Trevor as his residence, an ideal spot for the man who would oversee construction of the superbly elegant piers of Pontcysyllte. About the only difference between the two men concerned their native territory; Telford retained a lifelong love of Eskdale and Scotland, but Davidson made no secret of his low opinion of the country and its inhabitants. Instead, he fell in love with Wales and the Welsh girl he married.

By 1794 the company was letting contracts for the main-line sections between Ellesmere Port and Chester in the north, and Frankton Junction and Shrewsbury in the south. Both were relatively straightforward, through flattish country. But the thorny problem of how to drive the canal between Chester, Wrexham and Trevor, and carry it on over the two major obstacles in its path – the valleys of the Dee and the Ceiriog – was still unsolved. Then, in January of that fateful year, the first plans for an aqueduct at Pontcysyllte were presented to a meeting of the Management Committee at Ellesmere:

> the plan of the Aqueduct at Pontcysyllte with three arches as prepared by Mr William Turner of Whitchurch, Architect, shall be adopted by this Committee with such alterations as Mr Jessop shall communicate to Mr Telford, and that Mr Telford do prepare a specification and

The Telford Inn at Trevor, originally known as Scotch House (Geograph/Rob Farrow).

proper sections and working drawings to enable workmen to give in estimates.

Turner's plan was, unsurprisingly, the conventional one: to cross the Dee at low level with a flight of locks on both sides of the valley, with all the additional expense and water supply problems involved. It would also sacrifice precious height and the promise of a fast passage for narrowboats between Chirk Bank and Trevor. The new kid on the block, Thomas Telford, was immediately dissatisfied with the scheme and somehow managed to convince the committee to procrastinate. A fortnight later they withdrew the advertised specifications for Turner's aqueduct and granted Telford £100 to prepare detailed plans for an alternative design. Turner, jealous of Telford's success and growing influence with Jessop, resigned.

Two months later Telford made a hurried visit to London to show his revolutionary plans to Jessop, who approved them. Telford must have been extremely convincing, because Jessop had previously supported the low-level aqueduct. But the older man was probably relieved to give his young colleague a fair crack of the whip, at the same time settling one of the canal company's most intractable problems. Telford, at the very start of his canal career, had little to lose. His plans were advertised for tender in the Liverpool, Manchester, Birmingham and London newspapers, and in May 1794 James Varley of Colne in Lancashire was offered a contract to build Pontcysyllte's stone piers.

The bold recommendation of the canal's two senior engineers took the management committee by surprise, and they soon got cold feet. How could fully laden narrowboats possibly float across the Dee Valley at a height of 126 feet above the river, especially in a cast-iron trough? The bitter William Turner entered the fray and denounced Telford's design as impracticable; he then started claiming payment for time spent on his rejected plans. At this critical juncture it was Jessop's change of mind that made the committee change theirs; Telford's design would not have gone forward without the older man's blessing.

Cast iron is an engineering material which to us seems prosaically old-fashioned. But in those days it was state of the art. Telford had seen the wonderful iron bridge erected in Coalbrookdale 12 years earlier, and was a convert; the local ironmasters, who had a commercial interest in fast and efficient canal transport to serve the industries of north-east Wales, were among its greatest supporters. The stage was set for a daring application of the new material.

Of course Telford knew that James Brindley, the first of England's great canal engineers, had successfully built aqueducts across rivers in the 1760s. But he had done so at low level, supporting the immense weight of the navigable channel and its puddled clay bed on squat stone or brick arches. This was no way to cross a deep valley at high level; but in the summer of 1794 the Pontcysyllte proposal, although accepted, was still a castle in the air. Telford badly needed to demonstrate his ideas on a smaller scale to convince the sceptics that a cast-iron trough was practicable. And as if the gods were on his side, a few months later he was presented with an opportunity he could hardly refuse. As he wrote to his Langholm friend Andrew Little:

> I have been appointed Engineer to another Canal called the Shrewsbury Canal. It reaches from the Town of Shrewsbury to the Coalleries in the neighbourhood of the Wrekin at a place called Ketley where there are likewise great Iron Works … There are several Locks, two Aqueducts and a Tunnel underground for about half a mile. I have just recommended an Iron Aqueduct for the most considerable; it is approved and will be executed under my direction upon a principle entirely new and which I am endeavouring to establish with regard to the application of iron.

The 17-mile Shrewsbury Canal had also been authorised by Parliament in 1793. Its locks were originally designed to take trains of four 20-foot horse-hauled tub boats, transporting coal from local mines to Shrewsbury town. Civil engineer William Clowes and ironmaster William Reynolds were given joint responsibility, but Clowes died part-way through construction. Telford's six-year association with Shrewsbury as county surveyor, and the support of its 'rosy cheeked old country gentlemen', who presumably suggested him as the ideal successor, meant he was appointed sometime between November 1794 and March 1795. It may seem surprising that the proprietors of the Ellesmere Canal accepted this development, which clearly threatened distraction from their own project; but they probably saw it as a chance for their new engineer, undoubtedly talented but as yet unproven, to cut his cast-iron teeth at somebody else's expense.

Telford's cast-iron aqueduct at Longdon-on-Tern (Geograph/Chris Allen).

The iron trough was grafted onto brick abutments built by William Clowes (Geograph/John Horton).

One of Telford's first tasks was to re-engineer the aqueduct over the River Tern at Longdon-on-Tern, designed by Clowes but swept away by floods before completion. A lesser man would have recommended a conventional brick or stone replacement, but Telford's ingenuity and determination, plus the influence of William Reynolds, carried the day.

The River Tern rises near Stoke-on-Trent and meanders 30 miles southwest through Market Drayton, the small village of Longdon-on-Tern, and the National Trust's Attingham Park, before joining the River Severn. Its confluence with the mighty Severn occurs within a mile or so of Viroconium, where Telford had excavated the Roman bath house six years earlier, and within a few miles of Shrewsbury and Ironbridge. He must have been delighted by the opportunity to build the world's first substantial cast-iron aqueduct across a river in such familiar territory. With luck it would serve as a great advertisement for his plans at Pontcysyllte, 30 miles to the northwest.

His creation stands today, and must be the only Grade 1 listed building that straddles a river and finds itself marooned in the middle of a field. The 187-foot iron trough, 7 feet 6 inches wide and 4 feet 6 inches deep, was cast in sections at William Reynolds' Ketley ironworks and installed in 1796; it remained in service until the Shrewsbury Canal was abandoned in 1944.

Longdon-on-Tern was not Telford's only 'great preparation' for his subsequent triumph at Pontcysyllte. He also used cast iron at Chirk Aqueduct – another extremely valuable trial of the new material. Construction started in 1796 and was completed by 1801. Chirk followed almost at once on Longdon-on-Tern, and in effect formed the centrepiece of a trio that introduced the world to navigable iron aqueducts.

Our cruise along the Llangollen Canal has already introduced us to Chirk Aqueduct, one of Thomas Telford's finest structures. Prior to Pontcysyllte it was the tallest navigable aqueduct ever built, and is now Grade II listed in both England and Wales. On just about any other canal it would rank as the outstanding engineering feature. On passing along it in a narrowboat, or viewing it from the valley floor of the River Ceiriog, it seems to be constructed entirely of stone. But, as we all know, exterior appearances are often deceptive; and in this case mellow stone and a placid ribbon of water conceal a heart of iron.

Engineer and author Tom Rolt summarised the aqueduct's design in his biography of Thomas Telford, published in 1958:

> It consists of ten spans of forty feet each carrying the canal seventy feet above the waters of the Ceiriog. Here again, as at Pont Cysyllte, both Telford and Jessop realised that the orthodox method of carrying the canal in its puddled bed would impose an impossible load upon such lofty piers and there was a great deal of discussion between the two men as to the best form of construction to adopt. Telford finally designed a masonry structure orthodox in exterior form but within which cast-iron plates, flanged and bolted together, were used to form a continuous metal bed for the waterway. At the sides, these plates were

Looking north along Chirk Aqueduct, with the Victorian railway viaduct on the left and Chirk Tunnel in the background (Geograph/Roger Kidd).

Looking south along Chirk Aqueduct (Geograph/John M. Wheatley).

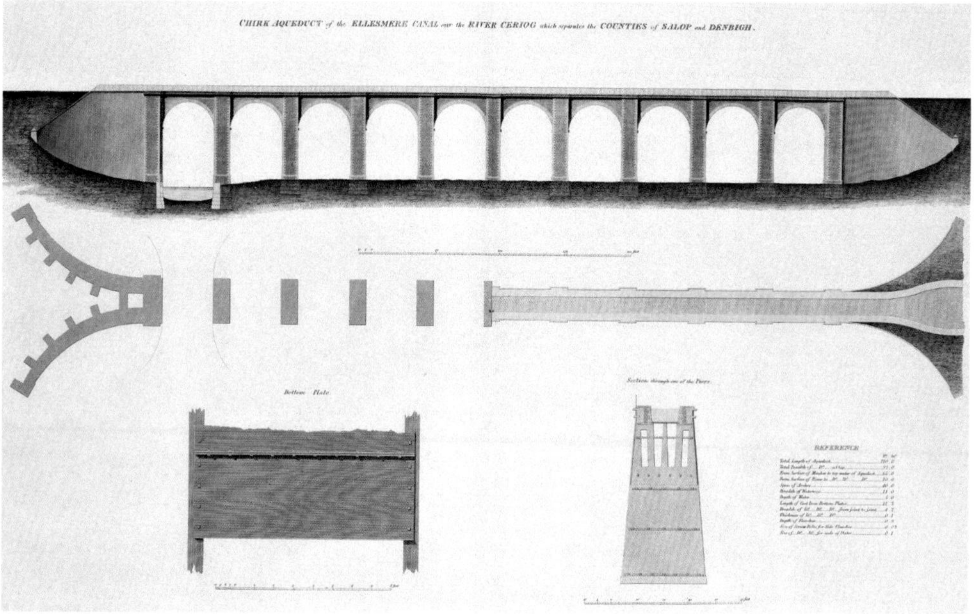

Engineering details of the Chirk Aqueduct (ICE).

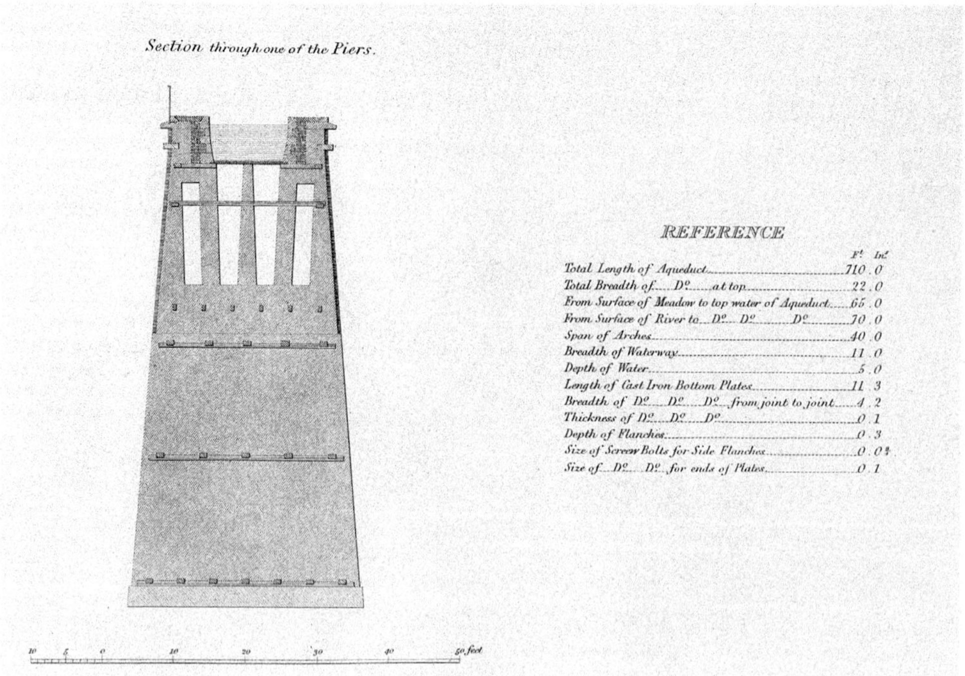

Enlarged view of a section through one of the piers. The table includes the following measurements: total length of aqueduct 710 feet; waterway 11 feet wide by 5 feet deep; cast-iron plates forming the bottom of the trough 11 feet 3 inches by 4 feet 2 inches, 1 inch thick (ICE).

securely keyed in ashlar masonry backed by brickwork set in Parker's cement. In this way the iron bottom not only formed a watertight bed for the canal which was much lighter and more efficient than clay puddle, it also firmly tied the side walls of the aqueduct together and so eliminated all risk of the blowouts due to lateral pressure of water and puddle which had bedevilled canal engineers in the past. For the tops of the piers and the spandrels of the arches Telford used the hollow wall technique with interior cross walls.

The accompanying drawings, taken from the *Atlas to the Life of Thomas Telford* published in 1838, show the structural details. The crucial bottom plates, hidden from view as soon as the trough was filled, were supplied by ironmaster William Hazledine, a friend and professional colleague who, as we shall see, figures prominently in the story of Pontcysyllte. The piers, far taller than anything Brindley had attempted, gave the overall design an elegance reminiscent of classical architecture. However, as with his canalside buildings, Telford preferred to regard the aqueduct as a symbol of the new canal age, and avoided unnecessary embellishments such as those used by John Rennie on the Kennet and Avon Canal near the neoclassical city of Bath.

Great preparations completed, the scene was set for an even more daring and spectacular application of the material that had fascinated Thomas Telford ever since, as a much younger man, he had marvelled at the Iron Bridge in Coalbrookdale.

Stream in the sky

The aqueduct of Pontcysylte [*sic*], over the River Dee, is one of the most stupendous works of art that was ever accomplished by man. The Ellesmere canal, which had run parallel with the farther bank of the river, is here carried over the vale by an aqueduct, supported on eighteen massy stone pillars … On the top of the pillars is a trough or water-way, wholly composed of plates of cast iron, about twelve feet in width and five and a half in depth. The inside, over which we walked, as the water was not yet admitted, is carefully pitched; and the manner in which the immense plates are connected and closed by ledges, bolts and screws on the outside, fills the mind with wonder as it contemplates the ingenuity of man. Mr Telford was the engineer; and if this were the only work he had produced, it would deservedly give immortality to his name.

This excerpt from William Mavor's *A tour in Wales and through Several Counties of England performed in the summer of 1805* declares Pontcysyllte 'stupendous'. Sir Walter Scott described it as the most impressive work of art he had ever seen. Percy Bysshe Shelley, one of England's best-loved romantic poets, called it a 'wonderful bridge of junctions'. Famous watercolourists, including John Varley and John Sell Cotman, made excursions to

Pontcysyllte in 1806: aquatint by John Parry (*Wikipedia*).

the valleys of the Dee and Ceiriog to record engineering structures that blended beautifully with, and even enhanced, dramatic landscapes.

Not that the admiration of 19th-century Britons was confined to poets, painters and aesthetes. In those days 'art' was seen to embrace a wide range of human activities. Utility and beauty complemented one another, and a fine aqueduct could perfectly well be described as a work of art. The Royal Society of Arts, founded in London in 1754 and granted a Royal Charter in 1847, existed 'for the Encouragement of Arts, Manufactures and Commerce' that benefited society, and numbered among its members such diverse characters as Adam Smith, Benjamin Franklin, Charles Dickens and William Hogarth. Engineering creativity in the service of man was considered as much an art as a science, and Pontcysyllte became widely acknowledged as a supreme example. Today, sadly, society as a whole seems to have lost the message.

Of course the physical expression of art rests on skill as well as vision. A civil engineer, particularly when engaged on a large-scale project, has a special responsibility for the safety of his creation and the people who will use it. Art without skill can be fatal.

After a long period of indecision William Jessop, having ditched William Turner's design, recommended crossing the valley of the Dee with a high-level aqueduct:

> It had been proposed to save expense in the Aqueduct at Pontcysylte [*sic*] to reduce the height 50 feet and descend and ascend by locks, but in due consideration I must now recommend to the Committee

> to make this saving by adopting an Iron Aqueduct at the full height originally intended which on correcting the Levels appears to be 125 feet above the surface of the water of the River Dee.

It all sounds clear and confident – but in the early summer of 1795 the design was far from finalised. Jessop expressed his anxieties in a confidential letter to Telford:

> In looking forward to the time when we shall be laying the Iron Trough on the Piers I foresee some difficulties that appear to be formidable – In the first place I see the men giddy and terrified in laying stones with such an immense depth underneath them with only a space of 6 feet wide & 10 feet long to stand upon, and the same want of room will hardly allow space for the Beams and scaffolding while the iron work is putting together.
>
> I therefore think in the first, that in order to reduce the weight of the Iron, or parts of it, it will be better to have the openings narrow by adding another Pier so as to have 8 openings of 52 feet from Centre to Centre instead of 60 feet.
>
> In the next place I would have the Piers 7 feet wide at the top instead of 6 feet, and make them about 2 feet more in the other dimension. I hope you will not have proceeded with the Foundations till you receive this. And I begin to think it may be better to have the canal 9 feet wide and 5 feet deep as was first intended.

The reference to 8 openings is surprising because we know that Pontcysyllte actually has 18 piers and 19 arches, with 53 feet between centres, giving a total length of 1,007 feet. It seems that even at this late stage Telford had intended his bridge aqueduct to stretch only from the steep northern slope of the valley to the far riverbank; he would cross the meadow south of the river on an extremely long embankment. What probably changed his mind, and swayed Jessop's, was progress on the aqueduct at Longdon-on-Tern, which suggested that an iron trough spanning most of the valley would be cheaper and more elegant.

Jessop's choice of a 9-foot wide waterway in the trough was also revisited. Telford figured that resistance to the passage of a fully-laden narrowboat would cause water to pile up at the bow and overflow the trough, so he increased the width of the waterway to 11 feet 10 inches, and cantilevered a 4-foot 8-inch towpath out from one side. This meant that as a boat moved forward the water it displaced would more easily escape beneath the towpath, unnoticed by man or horse, and stay within the confines of the trough.

After all the dithering and procrastination it was time for action. The foundation stone of the first pier was laid, as it might be today, by a local VIP, Mr Richard Myddelton of Chirk Castle; he would soon witness Chirk Aqueduct rising from the Ceiriog Valley in full view of his estate, and had to be kept onside. The ceremony took place on 25 July 1795, a month before details of the aqueduct were finalised. Matthew Davidson was entrusted with superintending the stonework, and in December the committee decided to build a house for him at the north end of the aqueduct. From now on Davidson had special responsibility for raising the stone piers that would support the 'stream in the sky'.

Amid all the action it is easy to forget that Telford had responsibilities along the whole route of the Ellesmere Canal. Pontcysyllte on its own would be enough to drive most of us crazy with worry and sleepless nights, but here he was, learning on the job, making decisions on everything from aqueducts to tunnels, embankments, cuttings and locks.

Yet even in this welter of professional activity, he found time for personal priorities. One of the most touching was a desire to revisit Eskdale and see his elderly mother who, according to Samuel Smiles, 'had descended far into the vale of years, and longed to see her son once more before she died'. Thomas had been relying on his friend Andrew Little for news of her, and had sent frequent sums of money to ensure she was well provisioned. But he feared that her indomitable spirit of independence, so well remembered from his childhood, would prevent her from acceptance. His plans for a visit were postponed for several months by an excessive workload, but he eventually reached Eskdale the following November, to find his mother 'alive, but that was all'.

On a lighter note, he called in on former Eskdale friends who found him the same 'canty fellow' as ever, and although risen greatly in the world 'not a bit set up'. Frank Beattie, a fellow stonemason who had turned his hand to village innkeeping, admitted that his mason's tools were 'all dispersed, perhaps lost'. Telford, however, had kept his: 'I have them all locked up in a room at Shrewsbury, as well as my old working clothes and leather apron: one can never tell what may happen.' Before saying goodbye to Langholm and his mother he refreshed his spirit among the heather hills of his beloved valley, and soon after revised his poem 'Eskdale', adding fresh touches and new lines. It was, however, only for private circulation among friends – not to be 'smuggled and sold'. In spite of increasing public exposure and recognition he was still, deep down, a private person.

By 1797 Pontcysyllte's stone piers were rising from valley floor and river bed under Matthew Davidson's expert eye. Telford had ordered samples of local stone to be tested for strength and weather-resistance. The two men agreed on a beautiful buff sandstone with specks of quartz that sparkled in the sunlight, obtained from a quarry at Cefn Mawr, just north of the aqueduct. For over two centuries the stone, fashioned to close tolerances by some of the country's finest stonemasons, has resisted corrosion by the elements, and cutwaters on the four piers that stand in the river have repelled fallen trees (and the occasional dead cow) washed down an angry Dee in times of flood.

Two of Pontcysyllte's cutwaters (© Crown copyright: RCAHMW).

Work on the slender piers was necessarily cautious and

Contemporary drawing showing piers rising from the riverbed, one of several levels of the temporary construction gangway and, at top left, men at work (© Crown copyright: RCAHMW).

painstaking. They were built solid for the first 70 feet and then, as at Chirk, hollow with interior cross walls. This saved masonry, reduced loading on the foundations, and allowed Davidson's expert eye to check the integrity of the stonework – no hidden joints, no rubble infill, no furtive 'scamping'. The piers, with outer walls just 2 feet thick at the top, would have to support a massive iron trough and 1,700 tons of water at a dizzy height above the river.

And dizzy it certainly was. Jessop knew it, and had recommended increasing the pier dimensions at the top to give a slightly larger working platform. Telford and Davidson knew it, and worried ceaselessly about the risk to life and limb. If you have walked across the aqueduct and felt slightly queasy in spite of the generous handrail, you may have felt it; and if you have travelled in a narrowboat along the trough, and gazed over the 6-inch rim – with no rail of any sort; just a sheer drop to the valley floor – you almost certainly have. Yet the challenge hardly compares with that faced by masons and labourers on a 10-foot × 12-foot platform 120 feet above the river, manipulating stones weighing up to half a ton.

Actually Telford tried to lessen the risk by acclimatising the men step-by-step:

> The stone piers are 18 in number, beside the two abutment piers; they were all built to the level of 20 feet, and then the scaffolding and gangways were all raised to that level; and the materials being brought from the north bank, the workmen always commenced at the most distant or south abutment pier, receding pier by pier to the north bank; and by this ascending from time to time in their work, they felt no more apprehension of danger when on the highest, than at first on the lowest gangways.

Which sounds a trifle optimistic, especially as he admits to a casualty:

> One man only fell during the whole of the operations in building the piers, and affixing the iron work upon their summit, and this took place from carelessness on his part.

Coming from a kind and thoughtful man, the comment surprises by its lack of compassion. He was presumably disappointed to have his safety record tarnished after all the care he had lavished on 'health and safety at work', a legislative concept unknown at the end of the 18th century.

The piers demanded great stonemasonry and construction skills but were essentially a modern application of an ancient craft. The Roman engineers who had built the Pont du Gard would have recognised it; so would the masons who had fashioned stones for England's historic cathedrals. But 1,007 feet of cast-iron trough was an entirely different matter: daringly innovative in conception, manufacture and installation, it required some very special talents. There were plenty of ironmasters in north-east Wales, Shropshire and further afield, watching with eagle eyes as Telford's plans unfolded.

Those plans included one more major engineering feat – the huge approach embankment jutting out from the south side of the Dee Valley to mesh seamlessly with the cast-iron trough. In 1949 Tom Rolt, with wife and friends, crossed the embankment aboard *Cressy*:

> Pontcysyllte aqueduct is approached by an embankment which is itself an unique feat of engineering, being no less than ninety-seven feet high at its tip. With the possible exception of the prehistoric Silbury Hill, no greater earthwork so far as I know had ever been raised in Britain by that date. Its slopes are now thickly clothed with trees which conceal its height and shut off the view of the valley below … These masking trees, however, do provide an effect of dramatic surprise, as I realised when "Cressy" slid out from their shade straight on to the towering aqueduct.

The massive earthwork was entrusted to William Davies, a local contractor. Two thousand feet long, it was a precursor to the mighty embankments raised in Britain's railway age. That it

William Hazledine at the age of 68 (after a pencil drawing by Francis Chantrey, 1831).

was constructed with tens of thousands of tons of spoil excavated from the Chirk and Whitehouses tunnels, and from the cutting between them, by an army of navvies with wheelbarrows, planks and shovels seems almost incredible.

William Hazledine (1763–1840), a friend of Thomas Telford, was among the ironmasters watching progress on the embankment and piers of Pontcysyllte. The son of a Shropshire millwright, and with relatives engaged in the iron trade, he had followed an apprenticeship with his father and gained valuable experience erecting corn mills and iron forges. In 1789, soon after Thomas Telford had moved to Shrewsbury, they met as members of the local Masonic Lodge and Telford began publicising Hazledine's work, especially his skills with iron. Hazledine soon became an ironmaster in his own right, building a new foundry at Shrewsbury and expanding it step by step to become one of the largest in the country, employing several hundred workers.

This remarkable man – 'Merlin Hazledine' as Telford nicknamed him – developed a huge talent for manufacturing and testing large iron castings. He used the latest machines, including James Watt's steam-powered beam engine, to produce leading-edge structural ironwork. In their time, Hazledine's factories were among the most important centres of bridge-building expertise in Britain. He and Telford would continue working together on many projects, including the world-famous Menai Suspension Bridge, completed in 1826 when Merlin was 63.

Progress on the Ellesmere Canal encouraged Hazledine to take a lease on the Plas Kynaston estate at Cefn Mawr, just a mile or so from Trevor Basin. He built a new foundry to export iron products along the canal, no doubt also hoping to land contracts for the Chirk and Pontcysyllte Aqueducts. As it turned out, the main line of the canal got no further north, but a short arm was built to Plas Kynaston to serve Hazledine's needs.

It seems that the Plas Kynaston furnace was in blast by 1799 because in that year Hazledine obtained the contract for supplying the iron plates for Chirk. In March 1802 the 39-year-old tendered successfully for the Pontcysyllte contract – offering to deliver the ironwork from Plas Kynaston and supervise erection of the trough above the River Dee. It secured him a pivotal role in the unfolding drama.

Actually, the letting of the trough contract was not without controversy. Established ironmasters John 'Iron-Mad' Wilkinson and William Reynolds had submitted a tender five years previously, but for some reason it was left to gather dust. This alone could have caused offence, but there were other issues. Reynolds had cooperated with Telford over the design and construction of the aqueduct at Longdon-on-Tern, and had encouraged his appointment by the Ellesmere Canal Company. But it seems that the company, realising that Hazledine

The southern approach embankment, now heavily covered by trees (© Crown copyright: RCAHMW).

The trough taking shape on top of the piers: watercolour by John Ingleby (© Crown copyright: RCAHMW).

Iron plates, flanges, nuts and bolts (Wikipedia/Arpingstone).

was busy setting up a new foundry at Plas Kynaston, decided to accept Telford's advice that they should wait, on the assumption that no other ironmaster could possibly compete on price 'delivered to site'. It may well have created bad blood between Telford and Reynolds, but was just one example of the engineer recommending, during a long career and as tactfully as possible, one contractor over another in an intensely competitive environment.

Erection of the cast-iron trough began in 1804 under William Hazledine's supervision. Constructed from iron plates 1 inch thick, its spans are 44 feet

Trough, towpath and handrail (Geograph/Chris Gunns).

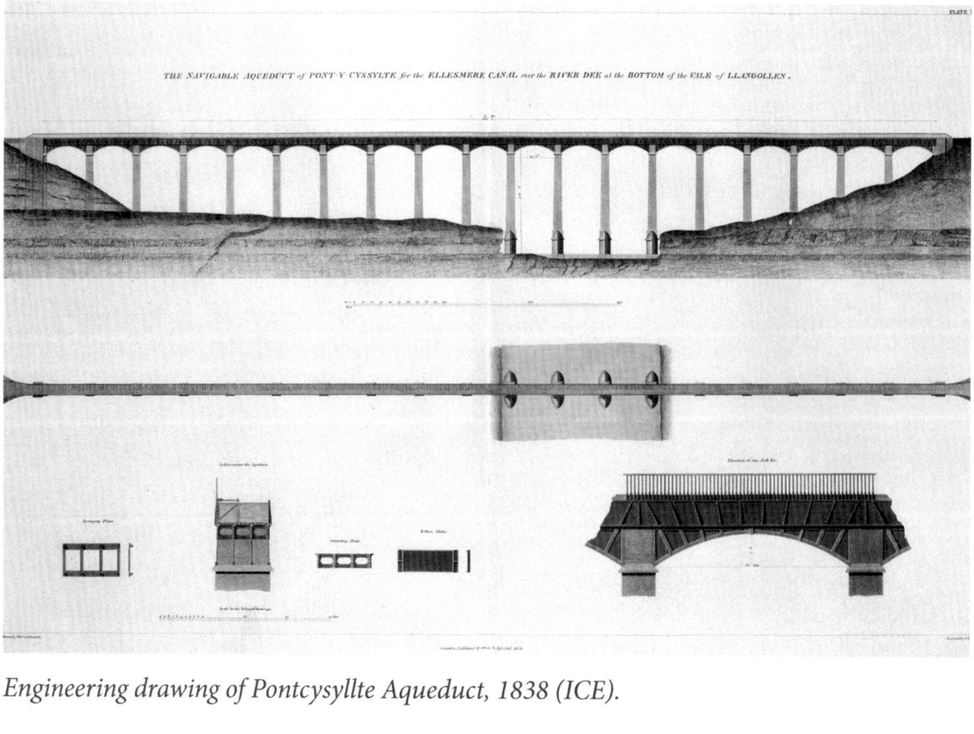

Engineering drawing of Pontcysyllte Aqueduct, 1838 (ICE).

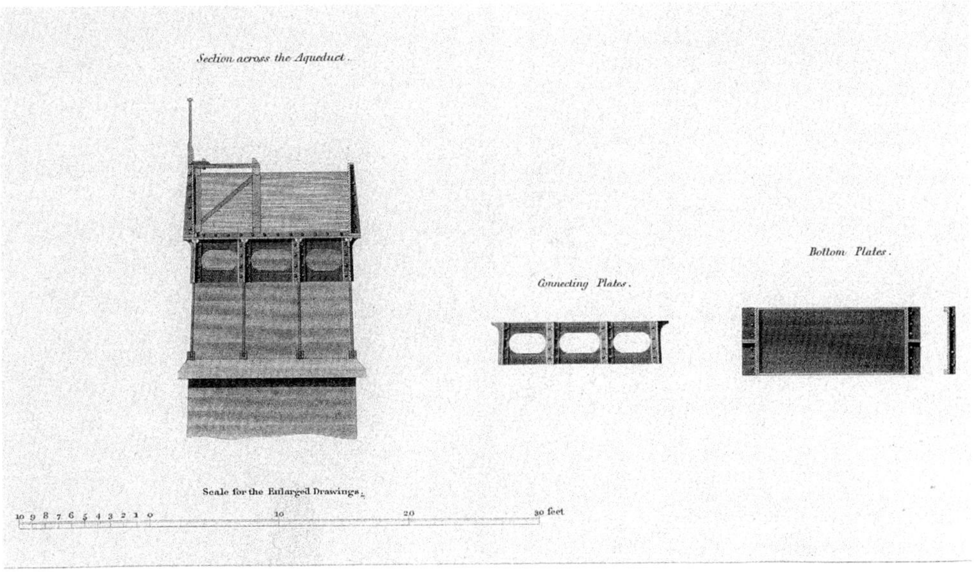

Enlarged view of the Ironwork: a section across the aqueduct; connecting plates; and bottom plates (ICE).

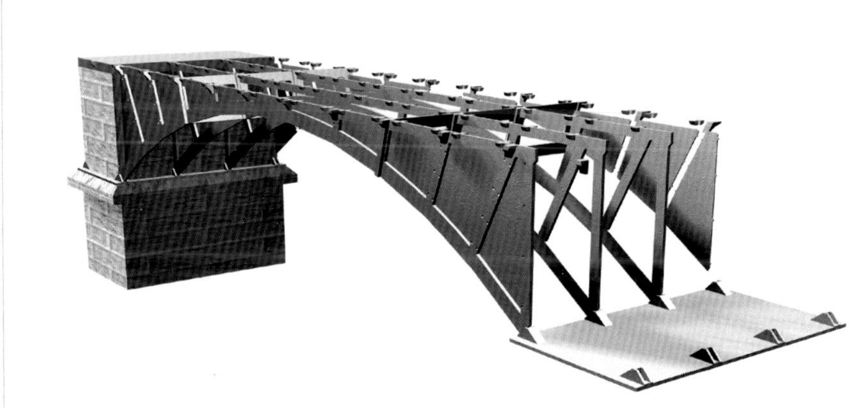

Four ribs resting on springing plates form the arch of each span (© Crown copyright: RCAHMW).

long and just under 12 feet wide. Supporting each span are four ribs, each made in three sections and bolted together to form an arch that rests on iron springing plates near the top of the piers. The two outer ribs have infill plates to give the impression of a solid arch, and the trough is prevented from lateral movement by brackets and lugs on its underside. Two of the accompanying drawings, published in 1838, show the completed aqueduct and ironwork. Note the enlarged 'section across the aqueduct' with water trough and cantilevered towpath, bottom plates and four supporting ribs. The third drawing is from a recent digital survey by the Royal Commission on the Ancient and Historical Monuments of Wales (RCAHMW), and gives a clearer view of the ribs and springing plates.

What next?

And so the construction phase of one of the most remarkable engineering structures in Britain came to a close. It is well named Pontcysyllte – the 'bridge that joins' in Welsh – as it strides triumphant across the valley of the Dee. Nothing like it had ever been imagined, let alone built, before; nothing quite like it would ever be built again. Its virtuosity and ambition are astounding. Tom Rolt, supremely qualified to comment on Britain's canal history, considered it 'undoubtedly Telford's masterpiece and the greatest single engineering work on Britain's waterways'. Few who cross it, on foot or narrowboat, would disagree.

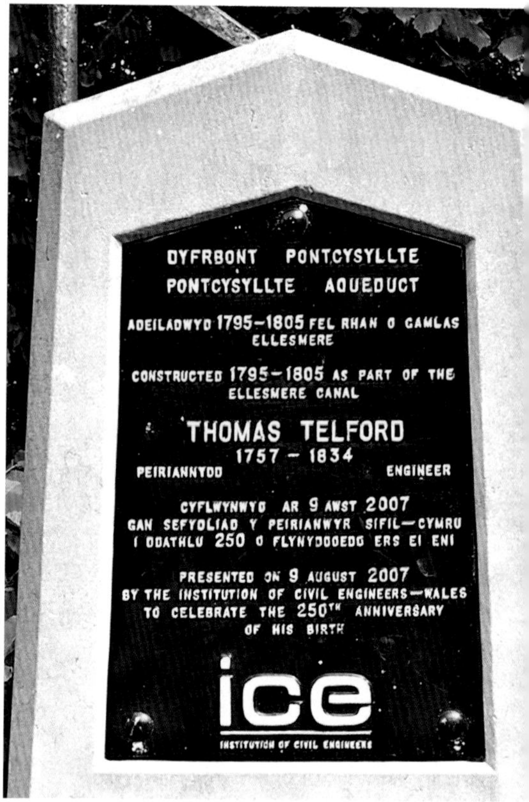

A plaque presented by the Institution of Civil Engineers in 2007 and placed at Pontcysyllte Aqueduct to celebrate the 250th anniversary of Telford's birth (Geograph/John Haynes).

The aqueduct has achieved international fame from its earliest days to the present. Writing in the *Wall Street Journal* in 2006, Hugh Pearman remarked:

> The structure and the landscape become as one, and the experience of moving slowly in a boat out from the shadows of the valley side and across the aqueduct, looking down at the Dee tumbling over rocks far below, approaches the sublime. Up there, so little separates water from air. Weight becomes weightless. Even at the time, it was acclaimed as a work of art as much as engineering … Two centuries on, it still functions exactly as intended. It is still awe-inspiring. It was thoroughly restored for its bicentennial, but it had stoically survived decade after decade of neglect in the 20th century. Looking at it today, it is as miraculous a fusion of architecture, engineering and nature as ever. This is a work of prodigious genius.

Chirk Aqueduct took four years to complete, but Pontcysyllte took ten (1795–1805). A lot of this was due to its novelty, size and complexity; but not all, because the canal mania gripping the nation's speculators was blowing itself out by 1794, to be replaced by the icy grip of a financial slump. Many plans for new canals were shelved, and much work in progress was questioned or frozen. So it is perhaps surprising that Pontcysyllte ever rose above the valley floor.

Although the Ellesmere Canal Company stuck with its two principal engineers and their remarkable aqueduct, it had little choice but to abandon its ambition for a main line stretching all the way from Shrewsbury to Ellesmere Port. The main victims were the collieries around Acrefair, north of Trevor, and the iron furnaces of Bersham and Brymbo, near Wrexham, major sources of wealth and trade that the northern part of the canal had been intended to serve.

From both financial and engineering viewpoints it is easy to see why the main line got no further north. The company simply could not afford the enormous cost of Jessop's plan for a 'great tunnel' beyond Trevor Basin; and probably not the alternative – a flight of locks climbing to a short summit level near Wrexham, and another descending to the Cheshire Plain. In any case there was no obvious water supply for a higher summit, and in spite of extensive surveying none was found. Jessop and Telford could hardly recommend two flights of thirsty locks, especially after engineering a level pound all the way from New Marton to Trevor, via Chirk and Pontcysyllte. It just didn't make engineering sense; nor, in a tightening financial situation, a commercial one.

At this awkward moment the Managing Committee racked its brains over alternatives. Was there another system that could raise and lower narrowboats without using so much water? Telford was ordered to investigate a vertical lift invented by two local gentlemen, consisting of a trough with watertight end doors into which a boat could be floated, attached to and counterbalanced by a 'diving chest' which rose and fell in a well sunk below the lower level of the canal. The unlikely contraption was built, but proved a dismal failure. The use of inclined planes – used on other canals with considerable success – was also considered and rejected.

Finally, all thoughts of connecting the aqueduct to Chester by water were abandoned. The only extension built north of Trevor Basin was a tramway to serve local collieries and ironworks, consisting of a double line of cast-iron rails. Needless to say, they were supplied, laid, and maintained by Merlin Hazledine of Plas Kynaston.

Of course this meant that a generous supply of water for the truncated Ellesmere Canal had to be found elsewhere. John Duncombe had previously surveyed a line for a possible navigable branch up the Vale of Llangollen to serve a local slate quarry, which would obtain its water from the River Dee at Llantysilio, beyond Llangollen town. The plan was now revived, revised and extended. The source of the Dee is in Snowdonia, and it feeds Bala Lake. Telford decided that the lake's capacity should be increased with a weir to allow more water to be released downstream in dry seasons. But the company needed consent from the lake's owner, Sir Watkin Williams Wynn, and Telford was given the tricky job of negotiating a settlement with him. Happily for a company facing the ignominy of a canal without water, his diplomacy bore fruit, and the plan was consented in June 1804.

In the meantime the southern end of the original main line, which started at Frankton Junction, had come to a stop near the village of Weston Lullingfields, about 9 miles short of the River Severn at Shrewsbury. The Ellesmere Canal, even if properly watered, was in danger of becoming a commercial millstone around the company's neck, cut off from the rest of the canal system – including its originally planned terminus, Ellesmere Port on the Mersey. Luckily the original plan had included a branch line northeast from Frankton through Ellesmere to Whitchurch, and this had progressed satisfactorily. The company now saw it as a relatively inexpensive solution: they could extend the branch from Whitchurch to join the Chester Canal (later part of the Shropshire Union) at Hurleston – the starting point of our own cruise – and abandon the plan to complete the main line beyond Weston Lullingfields. The work was pushed forward with all possible haste, and completed, like Pontcysyllte, in 1805.

So at long last, after machinations worthy of Machiavelli, the company had a water route between Pontcysyllte and Ellesmere Port, via Ellesmere and Hurleston Junction, which could serve, albeit in a very roundabout way, the industries of north-east Wales.

It is sometimes suggested that all this made Pontcysyllte little better than a stranded whale. I have even seen it described, in a recent book about Telford, as 'a largely purposeless curiosity for most of its existence', with the added comment that 'by the middle of the nineteenth century his life's work was a curious but useless dead end to the engineers who followed him'. But this is surely to misunderstand the nature of engineering science, and its practical application, on several levels.

For a start, Telford and Jessop were outstanding engineers, not money men or politicians. While it is true that Pontcysyllte never carried the weight of the commercial traffic it was intended for, this was due to the Ellesmere Canal Company's decision, following the collapse of canal mania, to amputate the canal's main line at both ends. Telford was in no way responsible for this state of affairs, nor could he have been expected to predict it; instead he designed a stunningly original and effective crossing of the Dee which, had the main line been completed as originally advertised, would have sped large numbers of fully laden narrowboats on their way in both directions.

Secondly, it is unjust to suggest that his life's work became 'curious but useless' soon after his death. Professional engineers know full well that applied science advances step by step, with the occasional quantum leap by a genius like Telford. Inventive engineers do not copy what has gone before, but they learn enormously from its successes – and its failures. Telford's willingness to take huge professional risks with cast iron advanced the knowledge of its possibilities and limitations, encouraging others to design iron railways and iron ships, and hastening the arrival of a modern steel industry.

Finally, we might reflect on the 19th-century view that an engineering masterpiece set in a dramatic landscape can perfectly well be described as a work of art, a thing of wonder as well as utility, an enduring tonic for the human spirit. Such things are never 'dead ends'.

Pontcysyllte certainly enjoyed a rapturous reception at the opening ceremony on 26 November 1805. According to Tom Rolt:

> It was a fine, calm afternoon when the procession of six boats swung away from Vron Cysyllte wharf, moved slowly along Davies's great approach embankment and then sailed out on to the high iron trough.

> The first two boats were occupied by members of the Managing Committee and their families, the next contained the band of the Shropshire Volunteers playing lustily, while the fourth was filled by the engineers, their families and staff. Bringing up the rear were two empty boats destined to carry the first loads of coal across the Dee. As this procession moved out over the aqueduct a mighty cheer went up from the thousands of spectators who crowded the riverside meadows or watched from the hill-tops, and the Vale of Llangollen echoed and re-echoed the thunder of cannon as a detachment of the Royal Artillery Company fired a fifteen-round salute.

But that is not all. As we might expect from an engineer who felt deeply about the natural world and man's place within it, Rolt's next words touch on matters spiritual as well as industrial:

> By the time the procession turned to recross the aqueduct and the cannons crashed out again the short day was nearly done; the sun had set behind the dark ruin of Glendower's hilltop fortress of Dinas Bran but as the November dusk thickened in the deep valley that wizard of the new world, "Merlin" Hazledine, coloured the scene with his magic as the false sunset glow of his furnaces flared and flickered skywards from Cefn Mawr. As the boats glided along, their occupants gazed down in awe from that vertiginous height upon the river rushing over its boulders, upon meadows veiled in rising mist and upon the pale moons of upturned faces far below. To those standing by the river gazing skyward at the boats as they moved slowly along their iron channel, dwarfed by the tall perspective of Telford's towering piers, the wonder of it was almost past belief. This was Pont Cysyllte, the magical stream in the sky.

Touchingly, but typically, it seems that Thomas Telford did not linger long at the site of his first great triumph. He had other things to do. So had the modest Jessop, who did not even turn up for the ceremony, but allowed his new colleague to take the plaudits. Neither was cut out to be a public showman.

Into the 21st Century

We may pass over the next 200 years of Pontcysyllte's history fairly quickly. Firstly, and pre-eminently, the aqueduct survived the vicissitudes of rivalry between privately owned canal companies, and the arrival of railways that eroded the canals' supremacy as a means of bulk transport. Already in severe decline between the two world wars, the canals enjoyed a brief respite between 1939 and 1945, when teams of young women crewed the narrowboats on important industrial routes, replacing men who had been called up. But the die was already cast: the canals were nationalised in 1947 and it was increasingly clear that if they were to have a future, it would be for leisure rather than trade. The Inland Waterways Association (IWA), formed in 1947 to campaign for their survival had, as one of its founder members, Tom Rolt.

We have already seen how Rolt and his wife tried to cruise the Llangollen Canal from Hurleston Junction to Pontcysyllte Aqueduct in their narrowboat *Cressy* in 1947. Three years previously the LMS Railway, owners of the canal, had obtained Parliamentary permission to close it. However, it was not in-filled or de-watered because of its vital role as a feeder for the Shropshire Union. This ensured the survival of Pontcysyllte as an aqueduct in the ancient sense – 'an artificial channel, conduit, or pipe that carries water from one place to another' – to which we might append the magic words 'iron trough'. By 1949 the canal had received minimal repairs to ward off danger and dereliction; the Rolts succeeded in their mission, and were delighted to find the aqueduct more or less unscathed.

Actually, Pontcysyllte had been little used by narrowboats over the previous half-century, and not at all since the 1930s. The owner's recipe could be summarised as adequate maintenance, minimal TLC. It was a magnificent survivor, almost unknown to the general public and disregarded by officialdom. But from now on it would benefit from a passionate campaign to save the moribund canal system, converting it to leisure use and rebirthing the Llangollen Canal as a jewel in the crown.

Erosion by wind, rain, snow and ice exacts a relentless toll on exposed structures. Lichens, mosses, weeds and even tree saplings can gain footholds in masonry and ironwork. And since Pontcysyllte was always a water feeder as well as a navigation channel, with a typical flow rate of about 2 miles per hour, debris and silt constantly entered from the northern end, to be trapped by flanges on the bottom plates. Churned up by horse-drawn boats, it caused further erosion.

Routine maintenance and minor repairs were no doubt carried out on Pontcysyllte from its earliest days. However modern narrowboats, driven by whirling propellers rather than horses, churn up water and silt far more violently, producing a shot-blasting effect on the inside of the trough. This accelerates corrosion of iron and damages protective coatings. More frequent remedial action is needed, planned for the winter months when canal traffic is at a minimum.

Benign neglect, softened by routine maintenance and minor repairs, can continue for a long time, but sooner or later something more drastic is required. In the winter of 2003/04 Pontcysyllte underwent a major restoration, costing about £1.3 million, which brought it to a state of perfection hardly known since 1805; a fitting tribute as its bicentenary approached –and, as things turned out, a key element in the successful application for UNESCO World Heritage status a few years later.

A lightweight aluminium scaffold was used for the restoration, underslung to give access below the trough, supporting ironwork, and masonry. It also gave weather protection to men working inside the trough and on the towpath. A cover of sheeting material, designed to rip open in wind speeds above 70 mph, allowed work to continue through the winter months in all weather conditions except severe gales.

Original tooling patterns on a pier (© Crown copyright: RCAHMW).

Replacing damaged stonework on one of the piers, beneath the iron trough (photo: Richard Turner-Thomas).

The only significant damage to the pier stonework was found where the iron springing plates are seated on the corbels. Water had entered the joints, expanded as ice in winter conditions, and dislodged pieces of stone. The problem had been noted some years previously, and repairs carried out, but unfortunately using stone obtained from Derbyshire. It was now decided to substitute local stone from Cefn Mawr and the original source, known as Vick's Quarry, was reopened. Most of the replacements were cut to size off-site, leaving only the outer face to be finished and tooled in situ. They were fixed with stainless steel dowels and epoxy resin bonding, and the joints pointed up with lime mortar where possible – but such was the workmanship that some of the joints were too tight to accept it. Telford would have been delighted.

Working inside the trough (photo: Richard Turner-Thomas).

Top-to-bottom inspection of the piers revealed tooling patterns so well preserved that individual stonemasons' work could easily be identified. However, some different patterns were observed at the ends of the aqueduct and, after much discussion, it was agreed that children sheltering under it over a 200-year period had thrown stones or objects at the inspection holes near the top of the piers, favoured by tawny owls and pipistrelle bats, and that some of the children being poor shots, their misguided missiles had scarred the surface of the stonework. But nothing could be done about it; nor about unwelcome graffiti accumulated over the years – better to leave well alone than provide a blank canvas for future 'artists'.

Three sets of stop planks were used to seal off the canal before the trough could be drained: one set at the entrance to Trevor Basin, full of narrowboats awaiting the 2004 season; one at the start of the navigable feeder to Llangollen, which had to be kept in water; and the third at the southern end of the aqueduct. The water supply right down to Hurleston Junction was maintained by a temporary pumping station installed on the south bank of the River Dee, close to the aqueduct.

The huge drainage plug on the aqueduct was opened on 24 November 2003, and by the following morning just a few inches of water remained in the trough. A fish rescue operation was mounted by a team of five men using an electric stun gun and nets to scoop up hundreds of roach, dace, bullheads, perch and a solitary ten-pound pike. All were transferred, apparently without complaint, to the Llangollen feeder. Then began the serious task of cleaning silt and debris from the trough, and carrying it away to the northern embankment in wheelbarrows. Working inside the trough, the men could ignore the 120-foot drop to the River Dee most of the time – but not when they climbed out onto the external scaffolding and faced the thrilling reality of their workplace.

Telford's design included four iron beams to support the trough over each bay, with fascia plates bolted onto the two outer beams to give the appearance of a solid arch. Many of the bolts had suffered corrosion, and some had sheared off. In one case just three bolts out of the original twelve remained to support a one-ton plate, and there was a worrying gap between plate and beam. Masonry work was immediately suspended. Had the plate fallen it would have smashed through the lower scaffolding and taken workers with it to an almost certain death.

Some of the beams showed cracks where they met the springing plates. Telford had been aware of this and had them monitored for some years after construction. All the cracks were now measured, recorded and photographed. In some places there was evidence of minor sideways movement, so steel tie bars were inserted between opposing fascia plates, spanning the full width of the trough and increasing the overall strength of the structure.

Once the inside of the trough had been thoroughly cleaned, the trough floor bolts – far larger and heavier than the fascia bolts – could be inspected for corrosion. The condemned floor bolts were removed using a specially made spanner with a handle a metre long. In all, some 750 bolts, complete with washers and nuts, and a further 1,000 nuts, were replaced by a firm of Derbyshire blacksmiths. The first discarded bolt was presented to the Aqueduct Inn at Froncysyllte. A pile soon began to grow at the northern end of the aqueduct, and then to shrink as local people heard about it: 200-year-old bolts, especially from a world-famous aqueduct, have a definite appeal, and today many local homes in Trevor and Froncysyllte are proud to display a trophy from 'the aqui', as it is affectionately known.

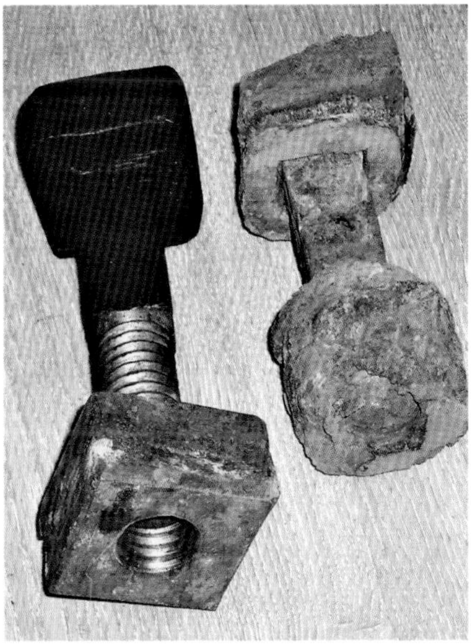

A severely corroded trough nut and bolt, and its specially made replacement (photo: Richard Turner-Thomas).

Finally, there was a major re-painting job. Shot blasting was rejected for removing old paint because it could have reduced the strength of iron components and released poisonous red lead into the pristine River Dee. So the men did it the hard way – by mechanical cleaning with power-operated wire brushes and hand-held paint scrapers. The trough floor was cleaned, dried and protected, using industrial vacuum cleaners and a gas flamethrower, followed by three coats of bitumen paint. For the first time in 200 years, Pontcysyllte had received a great deal of TLC.

Restoration almost complete, 8 March 2004 (photo: Richard Turner-Thomas).

Restored and repainted (© Crown copyright: RCAHMW).

Thomas Telford's 'stream in the sky' was reopened to narrowboat traffic on 12 March 2004 to great media acclaim. According to reliable reports a celebration party for the site workforce at a Llangollen hotel witnessed conviviality considerably more robust than anticipated. And so ended one of the most challenging restorations in canal history, a fitting tribute to the genius of its designer, the skills of Matthew Davidson and the iron-willed cooperation of Merlin Hazledine of Plas Kynaston.

6 World Heritage

> Situated in north-eastern Wales, the 18 kilometre long Pontcysyllte Aqueduct and Canal is a feat of civil engineering of the Industrial Revolution, completed in the early years of the 19th century. Covering a difficult geographical setting, the building of the canal required substantial, bold civil engineering solutions, especially as it was built without using locks. The aqueduct is a pioneering masterpiece of engineering and monumental metal architecture, conceived by the celebrated civil engineer Thomas Telford. The use of both cast and wrought iron in the aqueduct enabled the construction of arches that were light and strong, producing an overall effect that is both monumental and elegant. The property is inscribed as a masterpiece of creative genius, and as a remarkable synthesis of expertise already acquired in Europe. It is also recognized as an innovative ensemble that inspired many projects all over the world.
>
> (UNESCO World Heritage list)

In many ways it is remarkable that our favourite stretch of the Llangollen Canal and its crowning achievement, Pontcysyllte, managed to climb aboard UNESCO's international list of World Heritage sites in 2009. It shares the honour with such international superstars as the pyramids of Egypt, the Taj Mahal and the Statue of Liberty. In Britain it rubs shoulders with 30 other sites including the Tower of London, Canterbury Cathedral and the Forth Bridge. Yet the Llangollen Canal is not particularly well known in England or Scotland – except to narrowboat enthusiasts. Pontcysyllte, more or less unpronounceable to many, was locatable on a map by almost nobody prior to its World Heritage award.

All of which demonstrates the international status of Thomas Telford, the determination of a few people, including Tom Rolt, to rescue the decaying English canal system after World War II, and the hard work by many people to give Pontcysyllte its well-deserved restoration in 2003–4.

It is extremely difficult to persuade the United Nations Educational, Scientific, and Cultural Organization (UNESCO) to grant World Heritage status to a site. This was fully appreciated by the small army of local authorities, organisations and individual experts who worked together on a Nomination Document for presentation to UNESCO.

Superbly illustrated in full colour, 'Pontcysyllte Aqueduct and Canal' was jointly published by Wrexham County Borough Council and the Royal Commission on the Ancient and Historical Monuments of Wales (RCAHMW) in 2008. And, as we know, it proved eminently persuasive.

UNESCO's key requirement for granting World Heritage status is that a site represents 'Outstanding Universal Value' and meets at least one of ten specific criteria. In the case of Pontcysyllte Aqueduct and Canal the UNESCO committee acknowledged its Outstanding Universal Value in the following terms:

> The Pontcysyllte Canal is a remarkable example of the construction of a human-engineered waterway in a difficult geographical environment, at the end of the 18th century and the start of the 19th century. It required extensive and boldly conceived civil engineering works.
>
> The Pontcysyllte Aqueduct is a pioneering masterpiece of engineering and monumental architecture by the famous civil engineer Thomas Telford. It was constructed using metal arches supported by tall, slender masonry piers. They are early and outstanding examples of the innovations brought about by the Industrial Revolution in Britain, where they made decisive development in transport capacities possible.
>
> They bear witness to very substantial international interchanges and influences in the fields of inland waterways, civil engineering, land-use planning, and the application of iron in structural design.

An aerial view of Pontcysyllte from the southwest (© Crown copyright: RCAHMW).

Rather than meeting just one criterion, the Pontcysyllte Aqueduct and Canal site was judged to satisfy three:

> **Criterion 1**: 'to represent a masterpiece of human creative genius.'
>
> The Pontcysyllte Aqueduct is a highly innovative monumental civil engineering structure, made using metal arches supported by high, slender masonry piers. It is the first great masterpiece of the civil engineer Thomas Telford and formed the basis of his outstanding international reputation. It bears witness to the production capacities of the British ironmaking industry, which were unique at that time.
>
> **Criterion 2**: 'to exhibit an important interchange of human values, over a span of time or within a cultural area of the world, on developments in architecture or technology, monumental arts, town-planning or landscape design.'
>
> The intensive construction of canals in Great Britain, from the second half of the 18th century onwards, and that of the Pontcysyllte Canal in particular in a difficult region, bear witness to considerable technical interchanges and decisive progress in the design and construction of artificial waterways.
>
> **Criterion 4**: 'to be an outstanding example of a type of building, architectural or technological ensemble, or landscape which illustrates a significant stage in human history'.
>
> The Pontcysyllte Canal and its civil engineering structures bear witness to a crucial stage in the development of heavy cargo transport in order to further the Industrial Revolution. They are outstanding representatives of its new technical and monumental possibilities.

The three criteria were met 'with integrity and authenticity' because the site had not been degraded over the years, nor had the minor repairs carried out on Pontcysyllte to preserve its safety materially affected its engineering character or aesthetic value.

The UNESCO committee's deliberations were certainly comprehensive. Their recommendations for the future included a management plan for conservation; allowing for strong growth in visitor numbers and canal traffic; careful assessment of likely impacts of climate change; and prevention of industrial pollution. In short, they left no stone unturned.

Following the award visitor numbers and boat movements on the canal started to rise rapidly; local services including shops and accommodation expanded; road traffic and parking became an increasing concern. These are, of course, the inevitable consequences of a quantum leap from relative obscurity to worldwide fame – consequences which

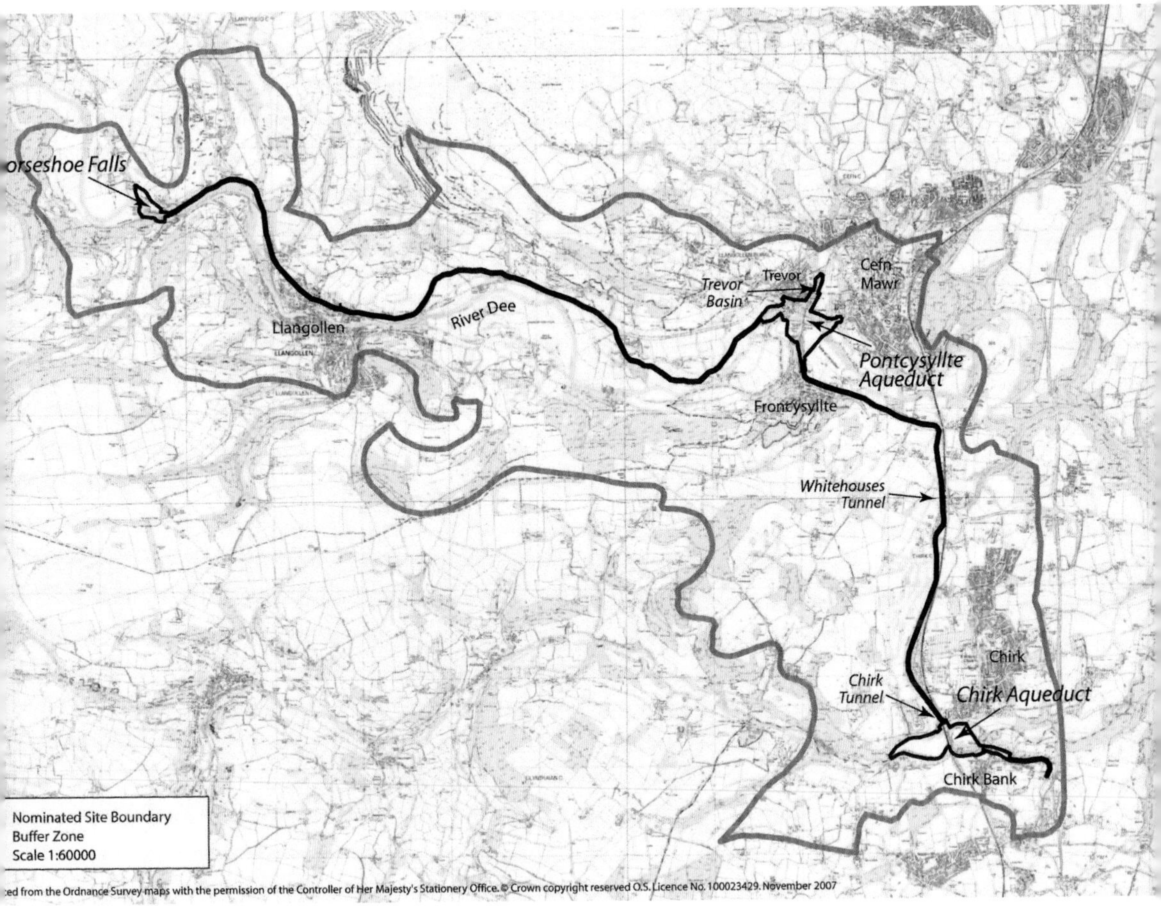

Map showing the 'nominated property' (dark blue line) and 'buffer zone' (within the dark green line). (© Crown copyright: RCAHMW).

local people have come to accept. World Heritage status has lifted Pontcysyllte and the Llangollen Canal into the stratosphere of internationally recognised tourist destinations, with all the benefits and adaptations that implies.

We might assume that the World Heritage site is restricted to the 11 miles (18 km) of canal plus its towpath and engineering structures. But in fact the 'nominated property' balloons out to encompass three additional areas, at Chirk Aqueduct, Pontcysyllte Aqueduct and Horseshoe Falls, as shown on the map. The total surface area of the site is 260 acres (105 hectares).

There is also a much larger 'buffer zone' of 16 square miles (4,145 hectares), extending for a distance on either side of the canal, in which development is carefully controlled to preserve the site's essential character and visual appeal. The aerial photograph gives a good idea of the landscape around Pontcysyllte, much of which is strikingly beautiful.

And how would Thomas Telford have reacted to all this? Probably delighted. He rarely sought fame, and never personal wealth, but he was passionate about engineering in the

Landscape and villages, looking northwest: Froncysyllte (bottom left); Pontcysyllte Aqueduct (centre); and Trevor (just beyond the aqueduct) (© Crown copyright: RCAHMW).

service of man and devoted his life to it. The best-known portrait of him, which hangs today in the Institution of Civil Engineers in London, depicts him as a famous engineer pondering the masterpiece that set him on the road to international fame – the wonderful Pontcysyllte.

Thomas Telford, President of the Institution of Civil Engineers between 1818 and 1834 (ICE).

Epilogue

Pontcysyllte Aqueduct marked the start of Telford's glittering 30-year career as a professional engineer. He went on to create other major works, including the Caledonian Canal, the Menai Suspension Bridge, the great road between London and North Wales, and the Gotha Canal in Sweden. He made an equally lasting but more humble impression – especially in his native Scotland – by designing a huge number of bridges, roads, canals, harbours and country churches, many of which survive today.

No man could possibly have 'built' such prolific works on his own. But Telford was a great judge of character and had the priceless knack of choosing, then supporting, able and conscientious colleagues who could carry out his original designs to the highest standards. Generous with praise for work well done but intolerant of the shoddy, he learned to combine the talents of a soloist with those of principal conductor.

Fortunately the 'Laughing Tam' personality of his youth stayed with him all his life. As Samuel Smiles put it:

> That Mr Telford was enabled to continue to so advanced an age employed on laborious and anxious work, was no doubt attributable in a great measure to the cheerfulness of his nature … When in society he gave himself up to it, and thoroughly enjoyed it. He did not sit apart, a moody and abstracted "lion"; nor desired to be regarded as "the great engineer", pondering new Menai Bridges. But he appeared in his natural character of a simple, intelligent, cheerful companion.

He was still working when he died in London on 2 September 1834 at the age of 77. Westminster Abbey offered him a final resting place and, close by, space for a larger-than-life marble statue showing him with a pair of dividers in one hand and two engineering books in the other. The inscription reads:

> Thomas Telford. President of the Institution of Civil Engineers: Born at Glendinning in Eskdale Dumfries-shire in MDCCLVII. Died in London MDCCCXXXIV. This marble has been erected near the spot where his remains are deposited by the friends who revered his virtues: but his noblest monuments are to be found amongst the great public works of this country. The orphan son of a shepherd, self educated, he raised himself by his extraordinary talents and integrity from the humble condition of an operative mason, and became one of the most eminent civil engineers of the age.

Index

Note: figures in bold type indicate an illustration.

Other highly-acclaimed books by Paul A. Lynn

- *...a gem of a book... ...Lynn blends art and science, distilling the story of Skerryvore into a highly readable, engrossing chronicle.* **The Lightkeeper**
- *The story of the lighthouse, and of the man who built it, Alan Stevenson, is a fascinating one, and Paul A.Lynn tells that story in an informative and highly accessible way.* **Undiscovered Scotland**

£16.99 978-184995-215-6

- *It's an inspiring history that's well worth exploring. Reading this book will certainly increase your appreciation of these buildings and the people who worked in and on them.* **60° North**
- *...a personal story of that dynasty of engineers ... Most ably the book places lives and work of Stevenson lighthouse engineers in their social and historical context and author Lynn writes of the Stevensons as a family of engineers.* **Flash**
- *There's lots of fascinating insight for both established fans of lighthouses and also those whose interest has just been taken by the subject matter.* **Best of British**

£16.99 978-184995-265-1

To order books please visit our website, www.whittlespublishing.com or call us on +44(0)1593 731333. We look forward to hearing from you.